PRAISE
THE SYNESTHESIA EXPERIENCE

"*The Synesthesia Experience* is not only the brilliant writing of a top, professional journalist looking in on a strange but romantic phenomena, but it is the writing of a person who could embrace the feelings of those she interviews, because author Seaberg herself possesses this remarkable gift of synesthesia. I predict when you pick up this book, you will be unable to put it down, as it will open up for you a whole new world in our universe."

—THE AMAZING KRESKIN

"As Maureen Seaberg beautifully tells us, the conscious minds of synesthetes reach deeper levels of reality, beneath the veil, beyond the cave, to purer realms of meaning."

— DR. STUART HAMEROFF, director, Center for Consciousness Studies, The University of Arizona, Tucson

"Maureen Seaberg explores a dimension of synesthesia long encountered in reports of synesthetes: its relation to mystical and artistic vision. In *The Synesthesia Experience*, Ms. Seaberg, a synesthete herself, has collected some fascinating accounts by some very prominent and inspiring people."

—PATRICIA LYNNE DUFFY, author of *Blue Cats and Chartreuse Kittens*

"If Husserl is right—that all perception is a gamble—then synesthete Ms. Seaberg is one of those gamblers who sits at the table with no one expecting her potency . . . then reveals herself as an artful, Zen monk with the entire Tao in her hand!"

—VANDA MIKOLOSKI, noted consciousness comedienne

"Maureen Seaberg goes beyond theory and abstraction to delve into the real-life experience of the (surprisingly) many whose senses have revealed themselves in ways to which the rest of us are oblivious. She helps liberate people to explore their own senses and discover the colors of sound, the sounds of color, and so much more. This is an exploration of the mysteries of the senses like no other."

—JAMES CLEMENT VAN PELT, program coordinator, Yale Divinity School Initiative in Religion, Science & Technology

"Synesthetes, particularly prominent ones, don't often share their perceptions and thoughts. By eliciting personal stories from famous synesthetic artists and scientists, Maureen Seaberg makes us aware of how people use synesthesia in their professional lives. Through these charming tales, she gives us new insights into the creative powers of synesthetic ways of perceiving and thinking."

—DR. CRETIEN VAN CAMPEN, psychologist and art historian, author of *The Hidden Sense*

"Synesthesia is no mere curiosity, but an important window onto human perception and creativity. This book's first-person contributions by figures such as Itzhak Perlman, Billy Joel, and Marian McPartland round out synesthesia's history and its place in the larger culture. Hand in hand with straightforward scientific accounts, these personal revelations speak to the meaning of the experience—for the individual, surely, but for the collective rest of us as well."

—RICHARD E. CYTOWIC, MD, George Washington University, author of *Wednesday Is Indigo Blue*

"The universe and all of existence is derived from quantum waves of possibilities. According to the research, those people who experience unusual manifestations of these possibilities in the form of the sound and light of synesthesia appear to be privileged by seeing more deeply into Creation than the rest of us."

—DR. AMIT GOSWAMI, quantum physicist, author of *The Self-Aware Universe*

THE SYNES

[TASTING WORDS]

THESIA

[SEEING MUSIC]

EXPER

[HEARING COLOR]

IENCE

MAUREEN SEABERG

ISBN: 978-1-63748-017-5
Library of Congress Cataloging-in-Publication Data

Names: Seaberg, Maureen Ann, author.
Title: The synesthesia experience : tasting words, seeing music, and
   hearing color / Maureen Seaberg.
Other titles: Tasting the universe
Description: Newburyport, MA : New Page Books, 2023. | Earlier edition
   published in 2011 as: Tasting the universe : people who see colors in
   words and rainbows in symphonies : a spiritual and scientific
   exploration of synesthesia. | Includes bibliographical references. |
   Summary: "Synesthesia is wondrous brain trait that is often described as
   blended senses. This book explores this fascinating subject, combining
   clear explanations of groundbreaking scientific research with an
   exploration of a deeper understanding of our senses"-- Provided by
   publisher.
Identifiers: LCCN 2023018840 | ISBN 9781637480175 (trade paperback) | ISBN
   9781633413214 (ebook)
Subjects: LCSH: Creative ability. | Synesthesia. | BISAC: SCIENCE / Life
   Sciences / Neuroscience | PSYCHOLOGY / Cognitive Neuroscience &
   Cognitive Neuropsychology
Classification: LCC BF408 .S3857 2023 | DDC 153.3/5--dc23/eng/20230501
LC record available at https://lccn.loc.gov/2023018840

Cover design by Sky Peck Design
Interior by Happenstance Type-O-Rama
Typeset in Ainslie Sans and Weiss Std

Printed in the United States of America
IBI
10 9 8 7 6 5 4 3 2 1

For my rainbow-hued tribe of synesthetes, particularly anyone who recognizes him- or herself in these pages for the first time.

CONTENTS

FOREWORD

Bob Dylan's song "Chimes of Freedom" was my first introduction to the phenomenon of synesthesia, and I didn't
even realize it at the time. At least, you could put it that
way. I was a teenager, around nineteen years old (and not
officially a synesthete), and I was listening to this song on
my record player (some of you may remember those), as I
had done many times before. But this time I was particularly
absorbed, and I had already started developing what I'll call
a literary or artistic consciousness. This time, I was suddenly
and unexpectedly gripped by the effects of the lyrics and
the music of the song for the first time. I experienced goose
bumps and sensations on the back of my neck, my heart rate
jumped, and I felt quite viscerally as though I were riding
a wave. When the song ended I went back and replayed
it a number of times, reexperiencing that aesthetic thrill. I
checked out the other songs/poems and found similar experiences with a number of them. I did not at the time attempt
to explain to myself what could have caused me to have
such a powerful experience.

After having this and other powerful experiences inspired by poetry and other art forms, I later came to understand that the intense poetic effectiveness of this song/poem came from a number of poetic techniques or "devices," including Dylan's extremely effective use of celestial and sky imagery (lightning, thunder, "wild ripping hail," clouds and "cracks" in the sky); his deliberately rapid and disjointed oscillation between ecstatic and terrifying apocalyptic scenarios; his use of the deep, ancient associations of bells (religious, mystical, royal, nuptial); his Dante-esque cataloging of human suffering; his collapsing and expanding of time and space; and, perhaps most importantly, his use of the powerful poetic device of literary synesthesia, the ecstatic blending of senses (or perhaps the psychotic breakdown of their normally rigid division)—in this case, of course, where bells cast shadows in sound, and where the "flashing" of chimes is simultaneously heard and "gazed upon."

I have long since given up the literary and poetic aspirations that the work of Bob Dylan and other poets inspired in me, leaving that to more capable hands, though I have entered a field that I consider to be no less important—the personal, scientific, and scholarly study of the world's "mystical" traditions, which has obvious connections to poetry and, as I have recently discovered, to the subject of synesthesia. I have been engaged for some years in the exploration of and investigation into the effects of meditation and related yogic practices (breath control, postural control, ascetic practice, etc.), especially on long-term, full-time advanced practitioners. More specifically, I have recently been focusing on the effects of such practices on the sensory-perceptual capacities of advanced adept or virtuoso practitioners, especially those in the Indo-Tibetan tradition. I have been incredibly fortunate to have collaborated with Professor Robert Thurman, His Holiness the Dalai Lama, and a number of leading scientists, scholars, and practitioners in this work. I am now further zeroing in on the claims by these traditions that certain meditative practices can lead to radically enhanced sensory-perceptual functioning. My orientation is an integrative, cross-cultural, interdisciplinary one, and I look within contemporary "Western" science (now actually "cosmopolitan" science, as it is an enterprise done by people from all over the world) and scholarship for possible bridges between science and these meditation traditions. I am beginning to discover some preliminary yet fascinating, extraordinary scientific support for such claims.

It is at this time, while affiliated with MIT and working at Tibet House in New York City, that my boss, Ganden Thurman—son of Professor Robert Thurman, brother of the extraordinary actress Uma, and longtime friend of the Dalai Lama—told me about his having recently been contacted by a friend and colleague, Maureen Seaberg, a freelance writer for the *New York Times* and other publications who was working on a book about the phenomenon of synesthesia. Ganden, though younger than I, has been like a mentor and guide in the world of Tibetan Buddhism and many other subjects, so when he emphasized the importance of working with this writer/researcher, the implicit high level of respect impressed me. I was aware that synesthesia was associated with my field of inquiry, but my awareness was not detailed or substantial. Nevertheless I offered to help in any way I could.

From the first time I spoke to Maureen (which lasted about three hours), I became aware that I was off on a new intellectual, artistic, scientific, and spiritual adventure of discovery. I was immediately inspired by her extraordinary enthusiasm and passion for this subject of inquiry and for all the subjects to which it is critically related: virtually all forms of artistic and literary endeavor; psychology and neuroscience; the philosophy of mind and consciousness; religion, mysticism, and spirituality; and even physics and quantum mechanics. During this period of time in which I have been privileged to collaborate with her on aspects of the project, I have been profoundly impressed time and again by the spectrum of areas of inquiry in which she has demonstrated not only wide-ranging knowledge but penetrating insight and the ability to tie together facts that, at first glance, appear disparate, even irrelevant and disconnected. In so doing, she has wound up inevitably "trespassing" on the borders of sectarian and "specialized" interests that characterize our present intellectual age, a time in which many seek to protect their narrowly focused domains by reflexive criticism and dismissal of those possessing an open-minded and open-hearted spirit of inquiry. With those rare few who were not won over by her rare gift of graciousness and genuine persuasiveness, Maureen has been unwilling to compromise regarding this interdisciplinary approach, and especially regarding what she considers to be the essentially transcendent nature of synesthesia.

Is it any wonder then, that Maureen's quest for knowledge about this unique brain gift takes her on a journey from darkness to light and from the mundane to the sublime? Go with her on her quest for understanding, prompted by her own "chimes of freedom flashing"—an epiphanic and light filled moment of clarity about synesthesia while writing of a dark and horrible crime. Participate, as I have, as she gets the cooperation of rock and rap stars, composers and painters, neuroscientists and physicists, lamas and Kabbalists, on her quest for the truth. Stand with her in the Guggenheim, listen to Itzhak Perlman's virtuoso performance with the New York Philharmonic, watch the dust devils swirl along the road in Tucson, Arizona, smell the fry bread at a Native American reservation, and kneel with her alongside Brahmins as she leaves no stone unturned in the search for answers. And perhaps most remarkably, close your eyes and imagine the beautiful and delicate sensations of synesthesia she and her fellow synesthetes painstakingly describe along the way.

Dr. William C. Bushell, PhD
Director of East-West Research for Tibet House

ACKNOWLEDGMENTS

There are many people without whom *The Synesthesia Experience* would still only be dreamy sketches on a coffee-stained tablet.

Thank you to my parents for a home that valued talking and telling stories and reading. Thank you to my husband for his support these many months. Thank you to my friends, particularly Lorraine Cancro, Siobhan O'Leary, Maria Alba Brunetti, Diane DiResta, Sandra Hunter, Lisa Marie Fricker, Shani Molligoda, Gul Celik, and Lauren Lawrence, for listening and inspiring and being the fine women they are.

To my agent, Jon Sternfeld of the Irene Goodman Agency, who never stopped believing, just as he once canoed the entire Mississippi River, and whose early edits shaped the book into something better than the ramblings of an inspired writer. To my acquiring editor, Michael Pye—thank you for seeing something in this project and for already knowing so much about synesthesia when my words landed on your desk.

To my editor, Kirsten Dalley, my "Northern Light," thank you for many days of further shaping and inspiring the manuscript, and a congeniality matched only by your erudition. Thank you especially to the Norman Mailer Writer's Colony, particularly my professor Dr. John Michael Lennon, President Lawrence Schiller, and my amazing colony mates for your early and continued guidance, support, and inspiration.

To the Tibet House founders and staff: Ganden Thurman, Dr. Robert Thurman, and Dr. Bill Bushell—eternal thanks for your light and wisdom. Thank you to the members of the American Synesthesia Association and other synesthesia community denizens, particularly Patricia Lynne Duffy, Carol Steen, Dr. Larry Marks, Dr. Richard Cytowic, Dr. David Eagleman, Dr. V. S. Ramachandran, Dr. Ed Hubbard, and Dr. Randolph Blake, as well as the many other synesthetes and researchers who shared their truths. And to the pioneers at the University of Arizona's Center for Consciousness Studies and their conference participants, particularly Abi Behar-Montefiore, Dr. Stuart Hameroff, Dr. David Chalmers, Cody Bahir, and James Clement van Pelt—I no longer have the zombie blues, thanks to you.

A heartfelt thank you to the many busy celebrities and their staffs who found time to talk with a stranger about their common rainbows, particularly Billy Joel, Itzhak Perlman, Pharrell Williams, The Amazing Kreskin, Ida Maria, Marian

Acknowledgments

McPartland, Dame Evelyn Glennie, Sir Robert Cailliau, and Dylan Lauren. You've never shone so beautifully.

Thank you to managing editor Angela Leroux-Lindsey and the staff at the *Adirondack Review* for believing in "The Red of His E String."

INTRODUCTION

I don't remember when the colors began. They've always been with me, like the beat of my own heart or the sound of my own breath. Science teaches that I was likely a fetus when my brain started forming the extra connections or began to have the lack of chemical inhibition that would enhance my world, creating a beautiful watercolor that only I could see. A Technicolor alphabet and numbers and days of the week, as well as colored months and music, would be my experience in life. I would have *synesthesia*, a blending of the senses, to go with my auburn hair, green eyes, left-handedness, and need for braces.

I believed this was normal for the first several years of life, as I learned the alphabet and numbers and how to use a calendar and play the piano. The internal wiring began to express itself in my *graphemes*, or symbols, as well as in the days of the week and names of the months. It didn't occur to me at first that other people didn't enjoy their letters the way I did, with all their swirling, attendant hues. That's the thing about life: If something's always been your reality, how do you know it's different for other people? When I eventually spoke of my impressions as a child, I quickly learned that my perceptions were not common; in fact, they were regarded as strange. Like many synesthetes around the world, I learned to keep them to myself. I'm grateful for the present-day climate of inquiry and wonder about this shy and deeply personal gift.

The experience of synesthesia is quite beautiful and, I believe, miraculous. There is evidence that meditative adepts, particularly those of the Tibetan and Zen Buddhist traditions, experience synesthesia in their meditations. Recently it has also been induced in hypnotized subjects. Synesthetes are also natural empaths who sometimes literally feel the pain of others due to the "mirror touch" neurons recently discovered in their brains. I noticed this as a reporter at accident scenes, and still do to this day, even when reading about people suffering in distant locales. (Too late I would learn that this also means that synesthetes should be very careful of their environments.) But all of these astonishing discoveries were made well into my adulthood. Like all synesthetes of my generation, I learned as a child not to speak of the strange, if ethereal and beautiful, tableau before my eyes.

As I write this as an adult and a journalist, the words I type still retain their spectrums. I can either choose to focus on the colors for each letter (which creates a lot of "noise" and slows my typing pace), or I can glance past them. Sometimes, when I go back to do a rewrite, whereas some people will correct alliteration, I will correct an over-representation of a certain color. I also sometimes substitute words, going to a thesaurus to find one that better "decorates" my phrases. It used to be frustrating when Microsoft Word underlined (in red!) the words *synesthesia* and *synesthetes* because it didn't recognize them. That is no longer the case and a sign of the newfound awareness of an interest in the traits.

I know the ink is black in my newspapers and books, but I see much more than that. Sometimes my colors radiate outward, in front of the actual characters, and sometimes they just create an inner recognition of each color in my mind. If you placed a sheet of cellophane on top of this page, with a different color for each of the letters, the result would be a pretty close approximation of what I see. I'd feel color-blind without it. There are definitely moments when it's more pronounced: I'll never forget my junior high school typing class and the letters spilling forth from the page like so many cans of luminous paint—teal Ks, terracotta Rs, indigo Hs, scarlet Es. Colored music often also dazzles me, and I recently discovered my motion-to-hearing synesthesia thanks to an online test. (The test showed balls of light streaming toward the center of the screen and then receding; as I looked at them, I heard the sound of rushing water, even though there was no audio.)

There are other challenges in navigating the world as a synesthete. I find sometimes that the color of an object is either perfect or distracting. For example, a Toyota Prius looks best in pale green or gray, but black really bothers me. Cash registers should not be industrial gray or beige, but fire-engine red. Dresses that drape a certain way should be one color and not another. And so on. I don't wish to impose my synesthesia aesthetic on others, however. Most synesthetes are grounded in reality, and I know that not everyone feels as I do.

My synesthesia is not usually an annoyance or a hindrance, however. It seldom overtakes the task at hand. I see it as additional information rather than a replacement for the main information I need to function properly. It is mostly a welcome background impression, a kind of music that is always with me. Anyone who likes multitasking or studying with music on in the background would understand that my synesthetic multitasking is not only possible, but very human. It is no different for me as a synesthete. It is just life, but it is life with a volume knob: It fades to the softest impressions when I need to concentrate on something else, but if I choose to focus on it, it can swell like the crescendos of a lovely movie soundtrack.

I lived in ignorance of what to call these impressions for a couple of decades—until the day I wandered, almost trancelike, into an Imaginarium store in a mall. I don't know what guided me in there, but it was just as though a hand were pushing me from behind. I was twenty-seven. There on a table, in the front left-hand side of the store next to the register, was a book whose multicolored title on a white background jumped out at me immediately. The title was *The Man Who Tasted Shapes*, and it was by Dr. Richard Cytowic. I intuited what it meant immediately. I read the inside flaps to be sure and walked the two short steps to the register to pay for it. I still get chills thinking about that moment of serendipity.

The next day I phoned Dr. Cytowic, the neuroscientist who helped White House Press Secretary James Brady in the wake of the attempted assassination of President Ronald Reagan. I wanted to do a story about him at my first newspaper job, a small daily. He was so charming; he gave me the interview and even told me about other synesthetes living in New York City. I set about to find these soul mates living in my community and polled a few local artists, intuitively suspecting that they also were "tribe members." I found one on the third try and included her in the report. But as a good newspaperwoman, I had to leave myself out of the mix.

Dr. Cytowic's book opened the floodgates. I snapped up several other engaging books, secretly harboring a desire to contribute my own. I would learn that more than one hundred years ago, synesthesia was not only known, it was *chic*. Illuminated color organs that paired notes with specific colors were once booming in auditoriums and salons; French symbolists such as Arthur Rimbaud and Charles Baudelaire were writing poetry replete with synesthetic impressions; and Charles Darwin's cousin, Sir Francis Galton, was studying people who had a unique ability to blend their senses.

Synesthesia's history may be lost to some, but it still plucks the strings of time in libraries around the world. There, you can learn that somewhere in a Weimer rehearsal hall, German composer Franz Liszt corrected his orchestra, "Please gentlemen, a little bluer if you please. This key demands it." Richard Wagner stormed off the stage during a rehearsal of *Tristan and Isolde* because in the maestro's eyes the set's colors were all wrong; they didn't match his expectations, given the music he'd written. (His royal benefactor agreed, and the opera opened with the new set.) Illuminated color organs, such as the one used for Scriabin's "Prometheus: Poem of Fire," once lit the stage and ignited imaginations, and the earlier yet futuristic 18th-century *clavecin de couleurs*, an illuminated harpsichord, caused audible gasps among corseted onlookers.

Despite art's fashionable love affair with the gift, synesthesia suddenly dropped off the radar around the time that psychiatrists began emphasizing behaviorism, and shying away from inner experience as a focus. It didn't return again until great modern researchers such as Dr. Larry Marks of Yale and Dr. Richard Cytowic took up the cause in recent decades. Now, modern brain imaging machines hum and buzz and gurgle out rainbow images of the brains of synesthetes, proving unequivocally that this is no artistic conceit. Synesthetes' brains *are* different: they light up in two sensory or conceptual areas when only one sense is stimulated. So as maddening as this 100-year-gap of interest in synesthesia is, new research is currently burgeoning. I take pride in learning that the gift is up to eight times more common in those in the arts, according to leading neuroscientist and author of *Phantoms of the Brain*, Dr. V. S. Ramachandran of the University of California at San Diego. Scientists are constantly discovering new types, and the diversity of its expression is evident in the synesthetes I spoke with in my research for this book. Synesthesia ends up being a large, colorful umbrella with great and varied wonders beneath it.

Although interest in synesthesia has ebbed and flowed over time, its genetic expression has endured. "It survived because it has a hidden agenda," Dr. Ramachandran has said, comparing it to how the sickle cell anemia gene protects people from malaria. "Synesthesia is about eight times more common among artists, writers, poets, and creative people. . . . If you assume that there is greater cross-wiring and concepts are located in different parts of the brain, then it's going to create a greater propensity toward metaphorical thinking and creativity in people with synesthesia," he said in a TED lecture. He later explained to me he would not go so far as to say the synesthesia gene is the creativity gene, but rather that it lays a foundation for creativity. "The reason it was preserved through evolution and natural selection is because it makes some people metaphorical and link ideas [and] be creative. You need this whole spectrum of human diversity."

Knowing that the experts have already had their say, it was beginning to occur to me that my quest should be to find the most prominent synesthetes in the world to talk about it. I'd done a lot of profiles on famous people throughout the years and witnessed firsthand their ability to bring a lot of attention to any topic they touched. A book that included the celebrated would certainly raise awareness of the gift. And though traditional medicine fascinated me, and research on the topic was finally picking up speed, I knew I must also call on consciousness experts. As someone who lives inside the experience, I knew that anatomy alone did not, could not, hold all the answers. In fact, many times, it didn't even know the right questions to ask.

There were beautiful moments along this journey. Dr. Stuart Hameroff invited members of the American Synesthesia Association to his conference in Tucson after we spoke about synesthesia and consciousness. While inviting me to hear her play at Lincoln Center one special night, jazz legend Marian McPartland admitted that she'd never heard the word *synesthesia* in her ninety-one years, even though each of her musical keys has a color. The great violinist Itzhak Perlman joked with me that the reason my Tuesdays are golden is because Monday is over. Tibet House medical anthropologist Dr. Bill Bushell, on a quest of his own to know the inner worlds of adept meditators, found synesthesia a fascinating aspect of that process, and a camaraderie grew from our parallel searches. Finally, doctors began to admit their own synesthetic moments and not only share their knowledge, but also talk about what was not yet known and left to find. The general goodwill shored me up to continue on my quest. It is a fabulous, creative tribe I feel fortunate to belong to.

—

This book seeks to explore and celebrate the experiences of many people who know the phenomenon as part of their lives and work. My hope is that this will inspire those with "different minds"—of all kinds—to embrace their uniqueness and use it in their creative endeavors. Perhaps it will help the scientists who now study the gift get more funding by popularizing the phenomena; perhaps a child in India experiencing it will one day know that the quintessential American songwriter Billy Joel thinks that certain music is azure, too. And perhaps it will help usher in a new artistic renaissance to accompany the strong scientific one that is now underway.

The book's tenor is not only scientific, but artistic and spiritual, as well. Multiple Grammy Award–winning producer Pharrell Williams believes that synesthesia is a communion with God. Dr. Bushell, a leading medical anthropologist, told me that ancient Buddhist texts imply there is no enlightenment without synesthesia. And Sir Robert Cailliau, the world-renowned engineer who settled on "www" as a name for the Internet because W is green on his synesthetic palette, says it's simply faulty wiring— an engineering glitch of the brain. But when he needs to remember something, his colors seem to act as a mnemonic device to retrieve information—a happy glitch, then. The beauty of the individual testimonies in this book is that they are the subjects' truths as only someone who knows the gift intimately can speak about it; many of them are speaking at length about it for the first time. Their words about such an ineffable experience helped me find my own. So the confused little girl has found community and some answers, but there are many more people out there still waiting, particularly children who may not have been born into sympathetic or knowledgeable families. But now, my quest has produced role models—people who've not only succeeded with synesthesia but who actually use it to inform their world-class art and their science. And for the first time, I advance the theory that synesthesia is nothing short of grace, evidenced by the Buddhist belief that it must be present at enlightenment. Indeed, according to two scientists I interview, the colorful forms that synesthetes see—the *photisms*—may actually be a form of quantum awareness. So there is much ground to cover.

Finding the right words to describe the ineffable has been the biggest part of my challenge. I'm reminded of Plato's allegory of the cave in *The Republic*, a work I first learned of in a political science class in high school. It had deep resonance for me because it reminded me of my challenges in describing synesthesia then, and now invokes that sense of coming out of a dark place and into the light that writing this book has engendered. In this allegory, the people chained inside the cave are only able to see the shadows cast upon the wall in front of them, from people passing in front of a fire behind them. The philosopher among them is freed and then tries to describe to the others the world outside the cave. What Plato is showing here is that the world of ideas, everything outside the cave, is truer than the world of shadows, or forms. And it is the philosopher who must transmit these truths.

To me, the photisms I see and the other impressions I receive are harbingers of a higher plane of reality than exists outside of this physical world of shadows. They feel pure and elemental and from the ultimate source, perhaps even quantum. So instead of describing a reality of trees and rivers and birds and other people—forms that the freed philosopher might have been inspired to bring to the shadow seers—I will speak to you of light and color and sound and feeling that are beyond most of your physical reality, yet which are, paradoxically, within all of us. And I will speak of the art and the *noesis* that result from this truer source. In the end I hope the shackles are unlocked and we are all standing outside the cave together.

1
–

THROUGH A GLASS, DARKLY

I'm a tiny girl. And I'm learning how to put the bright crayon colors on the first layer of the drawing and then color it over entirely in black. "You see," the teacher explains, "when you scratch the darkness off, you're left with the bright colors below. Take your pencils, and draw anything you want on the black crayon layer when you're done. You'll see it, the beautiful colors beneath." I take my pencil and press hard with my left hand and watch as the waxy swirls of black curl over on themselves or flake away. I write my own name.

I'm three years old. And I'm screaming in my antique cast iron bed at the top of my lungs in the middle of the night. It is 2:00 a.m. and I am sitting upright with both my little hands caught up to the wrists in the ornate bars at the foot of the bed. I couldn't have put them in there consciously; it hurts too much, it's too tight. And I'm terrified and confused about how I got this way.

My parents run down the hall from their bedroom, my father leaping over the safety gate. I'm inconsolable. Exhausted by my shrieking, I just want to lie down. Every time I fall over, I yank at my wrists inadvertently, then bolt upright from the pain and another wave of wailing begins. It must have been quite a sight, even for my father, a New York City police officer and former Marine accustomed to seeing people trapped and needing rescue. They work hard with soap and Vaseline. When they free me, I lie back down in the pitch blackness. As I try to fall asleep, my hands are throbbing. *Was it a dream about the jails my father had to put bad guys in? Had one of them wished it on me? Did a monster tiptoe into my room while I was sleeping?* I'm still confused. I comfort myself watching the colored forms in the air, my own private nursery mobile, which other people cannot see, as I later find out. These visions are normal to me. I don't know yet how rare they are. I follow them wherever they go, with wide, blue-green eyes. They are a balm in a childhood that is not worry-free; indeed, at times it's not really a childhood at all. But in those moments right before I fall asleep, I'm a blissful babe, still infused with the energy of this other realm from which I came.

It's the Summer of Love, 1969, and Joni Mitchell is sing-ing on the radio about stardust and being golden. I think it's not just Ms. Mitchell but everyone who sees what I do, particularly the teenagers in their tie-dyed T-shirts, whose colorful swirls look familiar to me. At that time I don't know most of their visions are chemically induced. I watch them on the news sometimes, swaying to and even swatting at what they think they see, and I giggle. I never swat at my visions; I'm used to them. The first few times I reach out for them I realize they can't be touched or held or moved, and they change and disappear quickly. And unlike the young adults I see on TV, I am not overwhelmed or disturbed by what I see. These visions are not part of the physical envi-ronment, that I am sure of. Though I can see something that no one else can, I'm still very much grounded in reality.

On television and in magazine advertising, the psyche-delia is everywhere. My parents are far from hippies, though; my policeman father has to work later than usual when they clash with law enforcement. We are "squares," and I can't be living in a more authoritarian or straighter household. And yet, here I am, this little girl tripping through visions natu-rally and thinking that the Zeitgeist is normal for everyone all of the time. There's never a reason to ask about what I am experiencing. I never really left the proverbial garden everyone was trying to get back to.

I have a brain gift it will take me twenty-five years to understand.

—

At this early age, despite the delicate and ethereal tableau before my eyes, I begin to exhibit a tendency to hurl myself, quite literally, into danger and chaos. Why someone gifted with this beauty—its teal, persimmon, magenta; its wonder—would choose darkness instead of embracing the wonder, is curious. Why someone would not feel blessed and proud, and instead want to keep these graceful spectrums a secret, is not at first understandable in this current age of greater interest in the gift. It is not long after the iron bars incident that I take an unannounced running leap over my father's lap, landing headfirst into the refrigerator grill and splitting my lip. Gravity and I will be coconspirators on many similar occasions going forward. I swing from a rope on a tree and miss a pile of leaves, breaking my arm. I hop fences and skin my knees so many times my father finally has to have a talk with me. "You won't like these scars when you're a grown woman. . . ." I like to move, and I wish I could fly like my father, over the edges of buildings on ropes sometimes.

"That cop was just like Superman," the *Daily News* quotes an observer on the street. The *New York Post* puts him on page one with our family dog, relaxing with a beer after another heroic exploit. My cousins in Virginia send up the front page of the *Washington Post* with an image of my father dragging a terrorist out of one of the missions to the

United Nations. My friend Lynne writes an essay for school comparing my father to the hero Beowulf, another giant Scandinavian.

At night I fly, too. When I dream, I leave my body and sail around my home or over the canopy of the plum tree in our yard, off to distant places. Often, it's the Big Apple, the great illuminated jungle gym across the bay that my father scales and swings from daily. I like soaring over the streets and dodging the spires like Wendy in a metropolitan Peter Pan. Sometimes a spirit guide appears on our street, which is named for the great Native American tribe, the Seneca, before I get too far. She manifests as a giant Raggedy Ann doll and catches me mid-air and redirects my flight, sending me sailing back into my body. I sometimes wake with a thud, landing back on my bed. As if the colored points of light I see are not strange enough, I have out-of-body experiences (OBEs) throughout my life. My dreams free me, and my experiences through them provide a sense of beauty that is lacking from the life my family is living, filled with the attendant stress of crime fighting and survival as it is. I have no idea then, nor do most scientists today, that these visions and the OBEs may be related. I will find that out much later, still.

I cry and cry when my two closest cousins go off to school first, because that means that I will have to wait one more year to solve the great mystery that is nagging at me, which is basically this: My letters and numbers seem tied somehow to color. Each one has a specific shade that rarely changes. And when I think of the numbers and letters they have their own colors, as well. Though I know that the symbols are usually black on a white background, to me they have additional "auras." When the school bus finally arrives one day, I plant myself next to the two older neighbor girls who told me they knew how to read. Perhaps they can unlock the mysteries of the colors I'm seeing (while failing to learn how to read, despite my Mom's best efforts) and that I can't stop staring at in their *Dick and Jane* books. I am looking for answers I won't find there. Somehow, however, I know it might not be a good idea to ask about the color part of the mystery; the girls don't say anything about this when they explain the sound each letter makes. My friends patiently share the reading part of the secret as we pronounce things aloud together from across the aisle in the back of the bus. What joy I feel the day I understand what the symbols mean, in spite of my lingering questions: *See the colors run. Run, colors, run!*

Despite my parents' guidance and sheltering, my father's exploits—hanging from pontoons of helicopters to rescue drowning men, climbing the Brooklyn Bridge to retrieve a bomb, or tethering himself to a daredevil who almost falls down the side of one of the towers of the World Trade Center—have me living in a state of constant fear and panic. I face what I am sure will be his inevitable death so many times that when the phone rings, I'm startled. I have two worlds: my inner sparkly one, and the bleak landscape of a terrified little daughter of a hero cop in an elite emergency unit.

There is little that isn't discussed about my father's days at our family dinner table, for we just read about it in the papers anyway, or learn it from the many journalists who phone or regularly show up at our home. Sometimes I even see it on the news before he's had a chance to call us. But despite the mysteries being solved around me daily from the city's crime blotter, I remain unsolved. One thing that is never discussed in our home are my visions. Like so many others of my generation who have mixed senses, I was born into an historical vacuum without knowledge of the phenomenon. I do not know that one hundred years ago, people actually knew more about it, and celebrated it in poetry and paintings and light shows at symphonies. Alas, I wasn't born into that *fin de siècle* celebration. So when most kids get around to questions like "Why is the sky blue?" I ask why the letter A is always yellow. My mother says I must have memorized it that way in school, but I hadn't. Well, why are my days of the week all colored, too? Why is Friday always black and Saturday always green? She answers that's probably due to the day-of-the-week underpants she bought me. But I know they don't match my personal colors at all. And where do my colored months and music and other experiences come from? I decide to let it go and stop asking. I feel strange, so why should I advertise the fact that my brain isn't working like everyone else's?

My mother's answers are actually very good ones at the time; no one knows or cares what naturally occurring synesthesia is in those days, despite a national fascination with the kind that comes from psychoactive drugs. I won't find out there is a name for it until I am twenty-seven years old and Dr. Richard Cytowic's groundbreaking book, *The Man Who Tasted Shapes,* comes out. In it, I learn that the colors around my letters and numbers are called "grapheme synesthesia." I learn that this is the "lower" type and that I also have the "higher" conceptual synesthesia, as evidenced in my colored days of the week, months, and music. Before long I am waving the book around and talking about it to anyone who will listen. But before all this, when I was growing up, my parents were doing all they could to shield me from the darkness of the world, despite the fact that the bad news from my father's life somehow managed to pierce my protective bubble on a daily basis. It was emotional triage— we had more to worry about than my idiosyncrasies.

At home, things are black and white, right or wrong, and there's not much room for subtlety. My mother quits her Wall Street secretarial job to raise me and my younger brother. She is concerned about my spiritual life, no doubt to balance my exposure to the city's worst stories. It helps me to know how to pray, as I know she does, over and over again for my father's safe return each shift. (Although she was raised a Catholic, she later decided not to give her children to that church when it refused to let her marry my father, a Lutheran, at the altar.) I also have rituals I perform.

In our first house, I pray to the *Apollo* astronauts, whose picture hangs on our kitchen wall opposite the black and white TV constantly tuned to the Vietnam War. They seem to fly farther than even my dad; these superheroes must have friends who are angels. And in our next house, I step on the same corner of the dining room carpet and stare at a decoupage of a London bobby's gear on the wall each time I pass. I imagine my father wearing the gear as some sort of protective device. It's my own version of magical thinking, I guess. I know from Sunday school that I should reach out to Jesus, and I do, but I think I've seen him and his girlfriend in Dr. Atlas's office being treated for hepatitis. I know because I walk up to him in the waiting room and ask him what he is in for, and make him repeat it a few times until I understand that it isn't "hippopotamus." I tell him I hope he feels better as my mom pulls me away. He sure looks like Jesus, anyway, so to hedge for Jesus possibly being very sick, I have these other charms.

I placate myself in later childhood by playing the piano. We have one in the sunroom and, self-taught and two-handed, I memorize the notes according to the colors of the letters. F, A, C, E are the notes between the lines—but for me they are pink, yellow, light blue, red. I play the *Muppet Movie* song "The Rainbow Connection" over and over, secretly assuring myself that they will one day (as the lyrics seem to promise me) find the rainbow connection, and it will have something to do with why Saturday and the number six and the letter S are all green, just like Kermit the frog.

The best thing of all is when my mom tries to teach me to do a jig in our kitchen on a rare carefree day. And it's not just her deft and light moves or her redheaded, pixyish good looks that tickle me, but the lyrics of the song, "The Orange and the Green." The Irish Rovers sing in a brogue I understand due to my maternal immigrant grandparents. I don't know then that the lyrics mean that the singer's father is Protestant and his mother Catholic, like mine. Instead I'm wondering how the songwriter chose his colors. To me, my father's name, Richard, is orangey-yellow, and my mother's name, Mary, is purple, so it doesn't quite match, but I like it anyway. I surmise that maybe his father's name is Richard but his mom's is Samantha, a green name to me. This is because words sometimes take on a color that has nothing to do with the individual colors of its letters; often the color of the first letter is the deciding factor. And I identify people by the colors of the first letters of their names as much as I would if they were blond or bald or tall.

My mother also imparts to me a love of literature and film. She reads to me often and lays the foundation that enables me to learn how to read on the school bus. She expects me to study hard, and I do. Every Sunday afternoon we watch old, black-and-white films on television together. She takes me to the Todt Hill-Westerleigh branch of the New York Public Library to get my first library card. I assume she hopes my unbounded energy might be channeled better this way. As we wander the stacks, my mother encourages me to pick out my first book: "Just reach for the one that looks interesting." I wander off and return with one on demonology. I'm fascinated by the winged creatures I see in there, mostly because this feeling of being scared is what's beginning to feel normal to me. I flip through the terrifying images and notice they even have names. My dear, Irish Catholic mother takes those dark mysterious creatures from my hands and marches me over to the Betsy books—a wholesome adventure series for little girls. I read all thirty in short order. I will need a lot more Sunday school and, Lord, am I about to get it.

Soon I'm in a Pennsylvania hollow attending a Mennonite vacation bible school near my grandparents' retirement home. I like the reading, the memory work. If you memorize a passage of scripture and recite it for the teachers, they affix a sticker on the inside cover of your Bible: merit badges for the Jesus set. I am very good at it and cover that red leather volume in no time, unaware that my synesthesia is probably responsible for my prodigious memory. "O taste and see that the Lord is good," from Psalms 34:8, is one that moves me the most. I accept unquestioningly the idea that someone could taste God in a cross-sensory experience. But it's the colors—the black Os, light yellow Ts, green Ss, and red Es—that really make the passage stick in my mind, even now, all these years later.

One day it is hot and hazy, and even the milky faces of the ladies in calico with hairnets look strained. The preacher at the front of the pavilion is shouting fire and brimstone, and I'm terrified I'm going to Hell. I'm passing in and out of consciousness until I black out. I wake to a sea of calico standing over and fanning me. I'm not sure where I am, and whether the colors of their garments are not my visions—*unless this is what these visions are like when you're dead*, I think to myself. The forms are opaque, dim, and stationary, not illuminated and translucent and motile. Finally, I see the crispness of the starched cotton beneath them, and I realize it's dresses, then women in dresses, and then the tunnel widens until I realize that I am at the center of this scene I have inadvertently created. Feeling ashamed, I shake it off as best I can and get to my feet.

Ultimately I decide to try to forget about my visions and just get on with things. And so, counter to my innate inquisitiveness, I do. It's odd that I don't push the issue as much as I tend to with other subjects, but later on I would learn that synesthesia testimonies are always like this. It's such an inner experience and such an unusual one, why would anyone risk disbelief, scorn, or worse? As a result, for most synesthetes, their gift becomes a very personal, very private, very secret experience. Part of the reason for this is that it defies description. The moment I write a word about it or say something out loud, I worry I haven't quite captured it. Even to describe it well sounds very odd indeed because, well, it is strange! And so I gasp but say nothing when, as a teenager, I see the mother ship land in Steven Spielberg's wondrous *Close Encounters of the Third Kind* and the scientists communicate with it via an illuminated colored screen, corresponding musical notes, and Kodály hand gestures. I dare not speak of how I understand these cross-sensory codes, lest I be considered freakish. And I am silent again later still, when I see *Flashdance* for the first time. In one scene, the Jennifer Beals character recalls how she was bored at her first symphony until she closed her eyes and "saw" the music, as her father instructed her to do. I feel such a bond with that character that I promptly cut off the shoulders and neckline of my favorite purple sweatshirt, the one that represents my initial, and I sport legwarmers. But I say nothing to anyone about the obvious synesthetic connection and how it resonates with me.

—

Eventually I realize that I want to be a hard news journalist, because they are the ones who know what's going on, even when I don't. There are times when I come home from school to find them sitting in our living room because my father has done something remarkable again, and they want more details while I'm still in the dark. I'd like to be omniscient like them someday, I think. I don't notice the lines on their faces or the antacids in their purses, of course; back then, they are my heroes.

A couple of decades later my assignments run the gamut from the mundane to murder and even the troubles in Ireland, traversing those orange and green lines that were so confusing to me as a child. I have a full and busy life and yet something is missing. I can't quite put my finger on it, but it feels a little bit like the "mean reds" that Holly Golightly talks about in *Breakfast at Tiffany's*. I definitely understand what Holly is feeling when she offers up this very synesthetic association written by Truman Capote, but the sadness and anxiety I'm feeling lately is more the color of bile. Emotions have colors to me, and I have a case of the mean yellow-greens.

Fast-forward. I'm sitting in the criminal court with the worst acoustics in the City of New York. Things stated here get swallowed in the high ceilings, which has the ironic effect of making you strain harder to hear things you'd probably rather not hear at all. That's never been more true than today. It's August, and the thick air further muffles the testimony. I'm struggling to take notes on the story of a horrific crime a man committed against his two-month-old infant son. My string is due in a couple of hours, whether or not I can hear well and whether or not I can stomach the proceedings, both of which are in doubt. I'm supposed to be objective but I'm furious—the first sign I should no longer be doing this. I realize as I'm sitting there that I've long had a familiarity with the dark side, if not a comfort with the subject matter. I relished the energy and the challenge of breaking news, perhaps exorcising some of the issues of my childhood by living that life in the first person instead of just hearing about it. But it no longer feels right for me. I no longer need to be immersed in the underbelly of society, but neither do I think that I should be writing about superficial topics, even if it's to save my sanity. I'm at a crossroads. I need to allow myself to think that there's beauty somewhere in this world. I'd like to go there and drink its restorative powers and write about it instead, but it seems like a journey light years away from this hard, wooden bench.

I've seen glimpses of it a few times—in the jasmine flower strands little street girls affix to their ponytails in Mumbai; in the ancient pastel-colored glass bottles in a Turkish museum that were once used to collect tears by the Romans; in the earnestness of a beautiful little boy named Obadiah living in the city projects; in a destroyed sand mandala carried off to New York City's lower bay by Tibetan monks; and yes, even in courtrooms. But those glimpses are fleeting. I must find my beauty elsewhere. As I sit there on the courtroom bench, I look at my wrists and find them stuck yet again in iron bars of my own choosing.

And I strain to think of one tiny beautiful thing.

And as I work to type it all later, I see it: tiny points of colored light around the letters I'm struggling to put down. Even as I type the terrible words that will carry this madness to the breakfast tables of people around the world, the ironic illuminated beauty of the colors of the letters belies their greater meaning. This now reminds me that we all come from light and eventually return to it. I hope against hope that this baby now has the light and peace he never had in life. Later the copyediting desk phones me to go over my words. An editor at the finest paper in the land, the *New York Times*, has the humanity to ask if I'm all right. "Just when you think you've seen everything," I muster the lame cliché. But what I'm thinking is that this must be what it feels like to stand in the midst of a total eclipse—not the benign eclipse of the still photos you see, with the corona of the sun still visible on the perimeter. No, it's more like I've been overcome by an Annie Dillard–esque moon shadow, seeing it swiftly race along the ground where I'm standing, earth-swallowing and final, the way she wrote in a short story once. I do a few more assignments but can't stop thinking about that little boy. Again, the two worlds: I chose the shadows in my youth and young adulthood, but I have come to realize that a life focused on beauty need not be without the mystery and excitement of discovery. The images that have been in front of my eyes for so long, but were overlooked as I hurled myself into blackness, suddenly become my focus once again. In fact, they pop against that void, brighter, fuller, and even more desirable than ever.

Will I be able to spin a new beauty from these threads in my mind? I decide I will live for a year as "synesthetically" as possible, seeking out and immersing myself in everything I can about the topic. I gather long-sought testimonies of likely synesthetes among prominent artists. I scour the events listings for exhibitions and concerts relevant to my quest. And, perhaps most importantly, I tell friends and colleagues what I'm seeking. They will fill my sails with inspiration and make some discoveries of their own along the way.

Will I be able to commit to mining my own imagination when for years I raced to events caused by others? I will find it's much harder to create something entirely new than take notes on someone else's narrative. This learning curve is just the challenge I need to face the page each day. I imagine a whole new set of words I can use now, replacing the ashen adjectives of my past with shimmering and light-filled descriptors of a wonderful mystery of the mind. As I write the book you are holding in your hands, I am as a small child again, lifting the pencil to scratch away decades of the soot and ash and finding these glorious spectrums beneath. Who will I be at the end of this task? Perhaps this is not the end of the rainbow, but only a beginning.

2
—

THE RED
OF HIS
E STRING

Where phones used to ring with news of the latest tragedy, my
house is now filled with the sweet strains of Itzhak Perlman's music.
He is teaching at his summer camp for children while I wait to
interview him. I imagine what it would be like to be sitting there in
his Long Island classroom, a young girl just aware of her colors.
I try to conjure the dappled light and the calm waters of the sound
and I think about how it is not far from where F. Scott Fitzgerald
once wrote so synesthetically of the "yellow cocktail music" that
played there during the Roaring Twenties. . . .

Inside a studio at the camp he runs for gifted young musicians on New York's Shelter Island, virtuoso violinist Itzhak Perlman keeps a huge jar of pretzels and a huge jar of jelly beans for the children to nibble on. It occurs to him a couple of days before speaking with me to open the jar of jelly beans and ask the students at the Perlman Music Program to each pick the bean that matches the color of the sound they are about to play. Mr. Perlman is relying on the jelly beans as an associative teaching device to get his young charges to see music more fully and from many sides. The class will then have to guess everyone's color. For Mr. Perlman, music has not just sound, but color, shape, and texture. In order to teach these young people to play more subtly and artfully, he wants them to imagine what color each note and tone might represent. They all choose different colors, which speaks to the great diversity of humankind and how music affects everyone uniquely. But they might not have known that for him, these are also sometimes literal associations. And this is the revelation that he has saved for me today.

I'm glad Mr. Perlman is thinking of me and our upcoming interview that day with the jelly beans. It hasn't been easy getting access to him; he's carefully protected by layers of people, as an international treasure should be. I spend many weeks sending letters and making phone calls to gatekeepers to confirm the longstanding rumor in the synesthesia community that he is a synesthete. A woman at Lincoln Center tells me she's been following him down the hallway there with the request and that he's thinking about it. I want to take my place among synesthetes and researchers and contribute to the history of and knowledge about this gift. Can he help me understand this mystery through his own experience? Just thinking of talking with him opens a shaft of light in my mind.

Finally, I prevail. He will speak to me when he has a break teaching at the prestigious camp for children he founded with his beautiful wife, Toby. When the day arrives, some of the adrenaline I knew at dangerous levels in my previous life as a breaking news reporter kicks in, but I notice it's a better kind of nervousness. When I hear his voice, I realize I can listen to his rich baritone for as long as his violin playing. It is like James Earl Jones and the voice of God rolled into one. His conversation is melodic, with bold, punctuating statements in places and sweet flourishes at other times, cadenced beautifully throughout and graced with a rich vocabulary. I suddenly realize why he is the finest violinist in the world: He is the violin.

Born in Israel in 1945, Itzhak Perlman studied the violin as a child at the Academy of Music in Tel Aviv. He appeared on *The Ed Sullivan Show* in 1958, which turned him into something of an international phenom. He went on to Juilliard, studying with Ivan Galamian and Dorothy DeLay, and then won the respected Leventritt Competition in 1964, which launched his career. In recent years he has also appeared as a conductor. President Reagan honored Mr. Perlman with a Medal of Liberty in 1986, and in December 2000, President Clinton awarded him the National Medal of Arts. And, of course, he serenaded the nation at the inauguration of President Obama. The legendary performer is a testament to the many abilities of the physically challenged, having been struck with polio at the age of four. Mr. Perlman is devoted to those with such differences, and it is that same gifted humanitarian who speaks to me now.

The humble maestro has never discussed his synesthetic associations with his musical friends, but he opens up readily when I ask him about it. "I know that I can describe certain sounds with color. It's not music—it's notes, it's single sounds. So if I hear a particular sound on a particular string on the violin I could associate that sound with color. . . . It's not like I play a piece and I see sparkling blue things." He begins slowly, tentatively. I, too, don't want people to think I walk around like someone on an acid trip seeing things all the time. I'm careful how I describe this, and he wants to be, too. But I understand that inner knowing, as well. Sometimes I don't "see" my synesthetic associations out in front of me or in my mind's eye, but rather "know" inside that, for example, Tuesday is golden. I want to share this so he won't feel so alone or strange or put on the spot. So I tell him about my inner golden Tuesday, and he laughs: "That's just because Monday is over." We are both laughing now. "If I play a B-flat on the G string, I would say that the color for me is probably deep forest green. And if I play an A on the E string, that would be red. If I play the next B, if I look at it right now, I would say that it's yellow. The bright colors are the upper strings of the violin. . . . I associate [that register] with bright colors of the spectrum." I try to isolate the individual notes in my mind; they dance in my mind's eye on a golden staff notation. I place them there for all time, in the same way that you might put a favorite memento in a glass case. When I'm recounting his story later, I instantaneously retrieve his colored notes from the staff, where they will live forever in a jumble of neurons that are either sitting too close to one another or lacking the chemical inhibition that is present in "normal" people.

To share one's personal associations is currency in the synesthesia world, and I am so grateful to know his personal palette. I think how my musical notes are imbued with color according to the letters that represent them; I certainly don't have a separate rainbow like his. Clearly there is great diversity within the synesthesia realm. In grouping bright or dark colors into high or low octaves, however, Mr. Perlman is like many synesthetes. According to Alan D. Baddelay in his book *Essentials of Human Memory*, a mild degree of synesthesia is very common: "Most people have a slight tendency to associate high-pitched sounds with bright colors and low-pitched sounds with more somber hues," he writes. And Bulat M. Galeyev noted it in his paper "The Nature and Functions of Synesthesia in Music," published in *Leonardo*, the MIT journal: "[Claude] Debussy relied upon the effect of that common synesthesia when he transposed the familiar motif in *The Lullaby of the Elephant Call* into the lowest register." In linking specific colors to the notes in those groups, however, the maestro is unique.

Mr. Perlman explains it in more detail: "One of the languages that one uses in teaching in describing what is missing from a [musical] phrase is, 'You need to give some color to these phrases,' or 'That phrase didn't have enough color,' or 'Change colors.'" He says that has to do with variety, and that shading is an even more precise description—certain colors can be less intense, bolder, and so on. And that's very much associated with the sound that one produces, he explains. When he teaches, he uses that analogy to describe what he wants. I realize it is common for musicians and even writers to talk about "coloring" a phrase. People without synesthesia can understand the sentiment implied by various colors: *Paint the town red, in a blue mood, green with envy, black-hearted, purple with rage.* Iconic writers from Emily Dickinson to Pablo Neruda have used cross-sensory pairings as metaphor. And yet, synesthesia is much more than that.

Dr. Oliver Sacks wrote about synesthesia and music in *Musicophilia: Tales of Music and the Brain.* "For most of us, the association of color and music is at the level of metaphor. 'Like' and 'as if' are the hallmarks of such metaphors. But for some people one sensory experience may instantly and automatically provoke another. For a true synesthete, there is no 'as if'—*simply an instant conjoining of sensations*" [author's emphasis]. Dr. Sacks gets at the heart of the matter here. If I wanted to be metaphorical about Mr. Perlman's friend Yo-Yo Ma playing Bach's *Prelude* from the first cello suite, I would pick something lovely, such as the sound made by spring's first flowers emerging from under the last winter snow (yes, there is such a sound!). In truth, though I love this piece and it makes me want to wax poetic as much as anyone else, I see coming from his strings a less bucolic image of ribbons of coffee in the shape of those glossy Christmas peppermints, the ones that fold back on themselves like waves. It is not a metaphor to me; it is a *real image*. And all synesthetes know the difference.

But this is not to say that synesthetes reject metaphor in favor of what they see or feel. In fact, they seem to have an aptitude for speaking and writing in metaphor, too. Dr. V. S. Ramachandran of the University of California at San Diego says that synesthesia is as much as eight times more common in creative people: poets, writers, musicians, and artists of various kinds. "If you assume that there is greater cross-wiring and concepts are located in different parts of the brain, then it's going to create a greater propensity toward metaphorical thinking and creativity in people with synesthesia," he said in a TED lecture. He told me he believes the gene for synesthesia is actually expressed more diffusely throughout synesthetes' brains than just the known hotspots, providing pathways that create an environment for linking seemingly unrelated things. If connected neurons live in concentrated groups in some minds, the synesthesia brain may have a wider "fishing net" of interconnected nerves.

So Mr. Perlman has both literal synesthesia and a propensity for metaphor, which is not surprising, as they seem to go together. And he says that his students (none of whom are synesthetes, incidentally) understand when he approaches music metaphorically. "I think that people get it in their own way. Everybody has a particular associ-ation with what you describe. . . . [U]nless you are really in the person's body, the way you describe what you want is extremely important. So language is really very, very important as to how you can say something to somebody and have them translate it in a particular way."

He doesn't remember how old he was when he first noticed his note-to-color associations. "I just felt that it was an obvious thing [. . .] because it's not like it's a gift that you can do tricks with or something like that. It's just something that you can associate with." But I think you can do tricks with it. I want to show him the beauty of the trait to repay him in some small way, so I ask if it ever comes in handy as a mnemonic device, given the huge amount of memoriza-tion he must do in his profession. For example, I ask him if he ever plays a piece of music and knows that "something red" is coming up? Amazingly, he says yes, that is the case! Now that he's really thinking about this hard-to-verbalize experience, Mr. Perlman has another revelation for me: "Besides colors, I see shapes," he admits. "Each note has a shape. I would say that if you play a D on the G string, for me that's round. But if you play an A on an E string, for me that's much more flat. . . . I hope not the intonation, but the shape of it." I feel a comfort level growing between us as he reveals the beautiful impressions he gets when playing his priceless 1714 Soil Stradivarius.

After hearing about this color-to-shape association from Mr. Perlman, I ask a top neuroscientist what's going on in the brain of someone who has this kind of synesthesia. Dr. David Eagleman, the author of *Sum: 40 Tales from the Afterlives*, has this to say: "The short answer is that the brain areas involved in color are next to—and in fact continuous with—the areas involved in form and texture. And this is why color synesthesia quite often involves more than just color." This also explains my "prelude in coffee" vision. Perhaps the note G lives in my brain next door to the color beige and Christmas peppermints. But if this were true, why do all the tests for synesthesia, including Dr. Eagleman's "Synesthesia Battery," rely on custom color bars alone, leaving the translation of the fuller experience wanting? "Interestingly, we cannot answer how rare it is to have more than color, because none of us have rigorously studied that so far," explains Dr. Eagleman. "And that's because it's easy to set up a color palette for consistency testing, but very difficult to set up an exhaustive form palette or texture palette to confirm synesthesia."

To Mr. Perlman, music both contains and expresses all of these things—shape, texture, form, color. In short, it's everything:

> It's everything that you can use to describe [a] note. The important thing is the way that it's helpful in identifying the sound, and in . . . being involved with the sound, to actually have a feel about what the sound does to you. It does to you with shapes, it does to you with intensity, it does to you with colors. And then you can really

*associate yourself with the sound. But sound—when you hear any-
body play an instrument or sing, the first thing that you hear, the
first thing that you're struck by, is [the] sound. You're struck by
the sound without even wanting to be struck by the sound. It's just
there; it's the first thing that meets your ear.*

He points out how if someone has an amazing sound on
an instrument—a resonant or rich or singing tone quality—
your ears perk up right away. And if someone does not have
a good sound, that's also immediately evident. This instan-
taneous judgment goes on within all of us in the blink of an
eye, before we even think the first conscious thought about
it. "This is the first thing that hits the listener, the sound,"
he explains. I think about how people's musical tastes vary
so much, but how there is still often a consensus of what
is beautiful. Perhaps there is another hidden, more universal quality to music that synesthesia provides "backstage"
access to.

Notes on a violin are not the only things that trigger the
color centers in Mr. Perlman's brain. He also associates color
with the human voice. He remembers somebody talking
about singers one day. They were talking about Pavarotti and
Domingo and other wonderful singers, and the sound quality
of the voice was the first thing that came up. Then someone
mentioned a third singer with whom Mr. Perlman was not as
well acquainted, and he later took the opportunity to sample
his music. "The voice [. . .] didn't catch you. The presence
was fine but not the quality of the voice. I remember this guy

and for me the sound of that voice was like yellowish-beige. I don't like yellowish-beige in my sound. I like, well, if you want to describe Pavarotti's sound, for me that is like metallic blue. It is amazing, there is a metal there, and I could describe it that way." He is right; Mr. Pavarotti had a voice like steel heated by the hottest flame.

Mr. Perlman says that his young students understand and access music through other means than just color associations. When he and another teacher were once talking about Brahms during a master class, the other teacher asked the students to describe the sound of Brahms. They didn't use color to describe it, but texture. "One word that came out was 'thick,' another word was 'heavy,' another word was 'full' and 'concentrated,' as opposed to something like Mendelssohn, which is airier. So there are a lot of ways to describe music, not just [with] colors." Those ways are as numerous as there are people playing music, he points out. And each of us is moved differently by the music we hear. When he talks with the children he teaches, he has one question that he always asks them: "I call it the 'goose-bumps moment.' In particular [what music] gives you a goose-bumps moment?" Although there is often a consensus regarding what is beautiful, this "goose-bumps moment" is different for everyone. It's all affected by what harmony does to the senses, he says. "And so I can listen to something Brahms wrote or Mozart or whomever it is, and absolutely start to cry."

He says he feels very lucky that he responds to music that way, and knows that not everyone does. "If you [see] ten people walking down the street, you cannot be guaranteed that everyone will listen to Brahms and be totally moved. So it's all very, very individual." He points out that people can be moved to tears at the age of four or five, well before most people develop a taste in music. They may not understand what they're listening to but react to it all the same. *What is that?* I wonder. Is it the music resonating with each person's own inner tuning fork? There have been times when I've cried while listening to music, and it wasn't always because of the lyrics. And how about the chills we all get sometimes? I think of Lara Fabian singing her Adagio in Italian and how I always shiver when I hear it, even though I'm not fluent in that beautiful language.

If there is a spiritual component to synesthesia or music, for Mr. Perlman it's in the vast diversity of the human experience: "There's a very personal aspect and everybody, I believe, is truly an individual. And each individual, [what] they do and how they hear, is totally individual. . . . [F]or example, how does harmony move you? What gives you a particular response? And why is it that when we hear certain things, we cry? That's not explainable." I remember Dr. Mehmet Oz telling me in an interview once how he plays recordings of the *ney* (the wooden flute used by the Sufi dervishes of his Turkish heritage) in his operating rooms because it actually helps stabilize surgery patients. Perhaps we don't play the music, but it plays us.

Mr. Perlman notes another magical moment in a teacher's experience: encountering the prodigy. "Sometimes when you hear somebody do something [and] the colors are so vivid, it's not necessarily that they've studied it; it's that it comes naturally," Mr. Perlman explains. "And that doesn't happen very often. But every now and then among kids, young kids sometimes, they do it naturally, and that you cannot teach. . . . If that's the case, then you don't have to do anything." He says that he has learned that great teaching is really all about allowing the good stuff to happen naturally and not over-teaching. He always says that the secret to being a very good teacher is not necessarily knowing what to say, but knowing what not to say. "You leave that gift alone. Words can be destructive or constructive, they can be both, and that's a fine line you have to be able to know." I wish every child could learn from Mr. Perlman. He is a careful gardener, and his words about this ineffable experience are helping me find my own.

Perhaps one of Mr. Perlman's most memorable performances was during the 1996 Academy Awards ceremony broadcast, when he played the "Theme from Schindler's List," composed by John Williams. He told the film critic Gene Shalit that he didn't hesitate for a moment when he was asked to bow the heartrending song, so realistic was its melody and so important its meaning. "I couldn't believe how authentic [Williams] got everything to sound and I said, 'John, where did it come from?'" Williams explained that he'd had some practice with *Fiddler on the Roof* and it all seemed to come naturally. "The subject of the movie was so important to me and I felt that I could contribute simply by just knowing the history and feeling the history, and indirectly, actually, being a victim of that history." When I ask Mr. Perlman what colors that now-classic composition elicits, he pauses. The song that so embodied the spirit of the film resonates with him still, but the associated emotion has created a synesthetic void. "I didn't think of colors there," he says quietly, reverently. I think about how the darkness of events can literally blot out one's synesthesia or make you overlook it; this certainly was the case with me for most of my life.

At the end of the interview I thank Mr. Perlman for many goose-bumps moments over the years—most importantly, the one I'm having now. "Very nice," he says softly, as the last pull of his voice's bow fades away at the at the end of our interlude. The red of his E string is like a long-awaited transfusion.

———

A few months later, Lincoln Center's Avery Fisher Hall is good enough to provide press seats for me for Mr. Perlman's performance with the New York Philharmonic, to aid my research. I watch with great wonder and joy as Mr. Perlman's bow seems to fly over the strings during Mendelssohn's violin concerto. His masterful playing is utterly mesmerizing, and watching his facial expressions change during various parts of the piece is an absolute delight. I feel less self-conscious about tapping my fingers on my knee to the rhythm when the couple next to me gasp out loud several times during the course of the concert. It's impossible not to react to the sounds we are hearing. When Mr. Perlman hits the impossibly high E toward the end of the movement, I feel goose bumps well from deep inside my abdomen and move up to my shoulders and arms, spilling tingles everywhere like a wave.

At intermission and the close of Mr. Perlman's portion of the performance, I go backstage, an honor and courtesy extended to me by Mr. Perlman's assistant, Mark Longo. There are dozens of well-wishers. When it is my turn, I reintroduce myself as "Maureen, the synesthete."

"Ah, yes," says the maestro. "Doesn't Mark look very blue today?" he jokes, a play on the color of Mr. Longo's shirt and the emotion, of course.

I say, "Yes, sir, but what color is blue on your Stradivarius?"

"F-sharp," he replies.

We speak in a fun synesthetic repartee that ends with his insistence that the bouquet of red roses I've brought to thank him is blue, because roses are blue. Of course I understand how a word's synesthetic color can differ from its literal meaning. "Well, yes," I play along. "I see that they are." And I don't exactly, as the word *rose* for me evokes swirls of terracotta, black, and red, because of the letters that comprise the actual word. But I understand how he might feel that way. Any non-synesthetes overhearing us would probably be scratching their heads, but I'm touched and gladdened by the cross-sensory shorthand between us. Learning from Mr. Perlman makes me more enthusiastic about continuing my quest to live with more awareness, synesthetically speaking, and to seek wisdom across the globe about these beautiful impressions. Instead of facing my days with trepidation, I find myself waking earlier, eager to turn the next corner on the greatest mystery I've ever written about. . . .

3

—

CAILLIAU'S GREAT GREEN WEB

It's May 2000, and I'm editor in chief of a website about India, traveling by rickshaw through the fetid yet beautiful streets of New Delhi, one hand always upon the huge duffel bag containing my cameras, which rattles and leaps each time we hit a pothole in the dusty byway. The old man stringing rosewood beads looks into my eyes as I pass his shop in a narrow alley, which opens onto a plaza. My driver pauses outside the Red Fort and tells me this is where they like to hold light shows as night, coordinating the

music to the colors of the lights. As elephants and taxis jockey for
positions in the foreground, I squint and try to imagine a perfect
powdery-blue middle C, bouncing from red clay edifice to red clay
edifice, up the drawbridge and through the scalloped passageways,
overtaking the cacophony that is this barely controlled chaos. I
think how the frontier of the Internet is like this street in India, full
of traffic and complexity and promise. I think about how we ride
the light as we help create this thing, and how marvelous the colors
look on the screen.

Sir Robert Cailliau has had to go through a secure keypad door or two in his time. He is one of the world's premier engineers, a developer of the World Wide Web, and a longtime employee at the European Organization for Nuclear Research (CERN), where a supercollider will one day answer questions not only about the origin of our universe, but life in the present day. He is also a synesthete. The Belgian-born man knighted by his homeland was recently faced with another of those guarded passageways. "A few weeks ago I had to go in and out of a building complex protected by a code that visitors had to type on a keypad at the entrance," he tells me. "I had memorized the code and things were fine. Then I had to go back there after a number of days and I could not quite remember the code," he recalls. Suddenly, the tiny particles that traverse the tunnel pathways of his own brain at high speeds got to work: His synesthesia kicked in. "Closing my eyes and thinking of it, I saw the color pattern and was thus able to fill in the digits I had forgotten."

I'm imagining the dashing Sir Robert in a scene worthy of *Mission Impossible*. Only it's even more thrilling to think that the solution to his predicament came from his own mind and not some movie-making magic device. Sir Robert, a PhD, is central to this discussion not only because he is a synesthete who has created a lot of wonderful content on the web for other synesthetes, but also because he is the man who worked with Sir Tim Berners-Lee of Great Britain in the actual creation of the World Wide Web. Initially meant as a communications medium for scientists working on the collider, his web is an achievement that altered the landscape for all of us, and, one might argue, for synesthetes in particular. We, who for so long had experienced an important aspect of ourselves largely in secret and silence, now have community as a result of this enormous contribution. Unlike when I was growing up, a great deal of really excellent information is now available online for people who are intrigued by their experiences and want to learn more. And there is real community now, where synesthetes can talk with one another on lists and forums and even Facebook groups. As a result, fewer and fewer synesthetes are living in silence. Therefore I've decided to reach out to him to discuss his own gift as well as the effect of the Internet on the synesthesia community.

How did it begin for Sir Robert? He first noticed that his numbers and letters had color in his mind's eye around the age of ten, he says. But he was almost fifty and had already changed the world before he knew there was a name for his special ability. "I was about ten or maybe younger," he recalls. "I do not remember [exactly] because I did not pay attention to the moment. I have only the common, 'mild,' form of seeing symbols in colors. People thought I was crazy, so I stopped talking about it. This is apparently a common phenomenon, ignoring it because others don't believe you anyway." When he was fifty or so, some people suggested that he had picked up the associations from children's alphabet books. "I recently found all three that I had in my parents' attic. I remembered only two, but none of the three had colors that even came close to my alphabet. So that could not have been it." It was only after 1990 that he came across an article in *New Scientist* that mentioned the trait by name. After he published his alphabet on his website he began to get a slew of emails from other people who had not realized there was a name for what they were experiencing—until now. "One woman thanked me because her husband now finally believes her!" he exclaimed. He told the website *mtbs3d.com* that he believes the web will bring more synesthetes to light: "[M]ost have been put off in early childhood by reactions of parents and friends—most are 'in the closet' so to speak, [and] it's only recently, mainly thanks to the web itself, that they have been able to understand their gift and make contact with other synesthetes. So I suspect we will see a higher number turn up."

One day when Sir Robert and his partner were relaxing with a beer on the cafeteria terrace after a hard day's work at CERN, Sir Tim Berners-Lee referred to the innovation as the

"World Wide Web." Sir Robert loved the green color that all those Ws elicited in his mind. Though he worried that the name might be a little cumbersome, the color impression won out. They went with it. Remember that this was in 1990, before there was any code written for the new invention. The tool whose name is now on the tip of everyone's tongue, and whose initials, "www," get typed countless times per day around the globe, is so named because a synesthete's favorite color is green! In fact, at an early conference about this new communications platform, green and white enamel "www" pins were given out to participants in recognition of his synesthetic association.

When I ask Sir Robert what synesthesia means to him, he responds with yet another computer analogy: "It has often worked as a 'negative' spelling checker (and it is 'spelling checker,' not 'spell checker'). When I see a spelling error I see a color clash, and while I may not instantly know the correct spelling, I still have been alerted." And here again is the beauty of having other synesthetes to talk to: I never realized it, but I've done the same thing when I've been stumped on a spelling. How did Sir Robert go from being a young man unable to speak of the inner life experience of synesthesia, to a man who would one day help provide answers for people like himself? He credits good parenting for his success. "The most influential were of course home environment, school, and contact with alert people. Your home and school environment [are] where, as a child, you can concentrate on certain things without having to worry about food, shelter, etc. You can just get on with what you want to explore. Yet you are also kept in line. Having caring parents is absolutely critical," he says.

He adds that a major part of the younger generation's problems today are neglectful or absent parenting. "There is a whole generation, beginning with mine, that was hijacked from parenting their children by pressures from commercial companies and the entertainment media. But advertising and the media [including the Internet, he says] are not yet pointed to in the same way as smoking is pointed to as a cause for lung cancer." In addition to good parenting, he believes he's been fortunate to have good people outside of his family around him throughout life. He directs me to his musings on his personal website: "At a few points in your life, you will be struck by a sentence, a happening, an action of a person, some sort of incident that is so at odds with the rest of your experience up to then, that you will never forget it. You had an insight. Germans call this an *'aha!-erlebnis,'* an 'ah-ha event.'" He keeps a list of such moments from his life, mostly in order to celebrate the wisdom he's gleaned from the good people he's had the good fortune to know.

Sir Robert says although nurture is the majority of his story, it is not the only factor. He became an engineer, he says, mostly because he likes exploring, building things, and tinkering. And though he is moved by meaningful events in his own life, Sir Robert doesn't read spirituality into his synesthesia. He defines spirituality as "just a side effect of our brains. That's not to say it is bad, just [. . .] that it is not [within] the scope of investigation. I purposefully avoid the adjective 'scientific,' because I see no reason for making a distinction between [science] and everyday decision-making."

Communicating about science is obviously important to Sir Robert. It was the topic of a paper he submitted to a European conference in 2008, and it will continue to be his goal in retirement (he left CERN in 2007 after thirty-three years of service). He wants to help bridge the gap between laypeople and scientists—something his Internet has been instrumental in accomplishing, as well. "The problems of communicating science stem from the non-human nature of scientific knowledge," he wrote. "Scientists must take this special nature into account and adapt their explanations to the psychology of a human public." He also calls for people to engage in lifelong learning as the world becomes more complex: "A familiarity with technology and the scientific nature of subtle inter-relationships is essential. Ideally, the public should have a good grasp of basic science. Communication of science is highly necessary and needs no defending. After all, there is something called the 'right to knowledge,' about which John Adams already wrote in 1765. It is mentioned as one of the cornerstones of the development of the European Union and the vision of the information society based on knowledge." He adds that the right to knowledge is a fundamental one that was stated in the Declaration of Human Rights. He believes that scientists have a responsibility to transmit their basic knowledge to people throughout the world.

Another thing he really enjoys talking about is what's going on at CERN. I'm happy that he shares news of the fascinating high-speed tunnel with me. "People are excited at the prospect of running the repaired machine after it broke down during final tests [in 2008]," he tells me. "But let us not forget these aspects: First, there have been large accelerators with large experiments before. We know well that results do not come in immediately and that they are usually far more complex to understand than was first thought. So a lot of work will have to be done before you will see any announcements of discoveries." Secondly, he says, the experiments have been planned for years. "The web is in fact the first spin-off of the Large Hadron Collider (LHC) era. It was meant to facilitate the publishing of documents related to the collider and to the experiments. In fact, the implementation of password access to websites was done in 1993 to ensure that there was some measure of privacy between groups in competing LHC experiments, who did not necessarily want to share the results of brainstorming sessions. Things take time." And finally, he explains, the results of the experiments will not only shed light on the origins of the universe, but will also help explain current and future events: "And I avoid here another adjective, 'physics.' Those phenomena are just as real as anything else happening around you; they are not 'physics phenomena,' happening in some other realm [and] to be ignored."

Just as Sir Robert hopes for a better general knowledge base about science, he is happy to see a greater awareness of synesthesia than existed previously. He points to his Internet as a key to this awareness. "I think the web has certainly raised the awareness. The web has brought together quite [a few] sparse groups, like people with special hobbies, with rare diseases, etc. In the first article I read it was estimated that only one in 25,000 people had synesthesia. But the web made [many people] more aware of their trait, and I think the [level of this awareness] now is higher."

Dr. Sean Day is a professor of anthropology and linguistics at Trident Technical College—and a synesthete. He also founded the most popular mailing list for the gift on the web. Known for his signature crisp purple dress shirts, he is also the president of the American Synesthesia Association. Dr. Day says that Sir Robert's invention has revolutionized research and community for his colorful peers. "The web has definitely changed the landscape for synesthetes, and has been critically instrumental in creating not only new, 'virtual' communities on the Internet, but also actual, real synesthete communities in various geographic areas." The Synesthesia List, which he founded in 1992, now has more than 1,250 members from fifty-two different countries, and is still growing. "I have no precise numbers on how many of those members are synesthetes, but I would roughly estimate that over half are. The other half is comprised of researchers, non-synesthete family members, artists, writers, journalists, and other interested individuals. Of course, there is also some overlap, such as members who are both a synesthete and researcher."

The Synesthesia List is not just for researchers and synesthetes or helping them connect with one another and share data; it is for *all* interested parties, including students, artists, writers, and so on. "The list is one of the few, and one of the first, [online] forums where neuroscience researchers and those with the neurological condition the researchers specialize in can directly interact with each other and ask each other questions," says Dr. Day. The main language for the List is currently English, but there are also postings in French, German, Italian, Russian, and Spanish. Email has also helped facilitate conferences related to synesthesia, he points out: "Via email, researchers and synesthetes from all over the world can coordinate meetings and academic pursuits." The web has also helped foster the formation of real-life groups, Dr. Day says. Several international conferences have been held and more are being organized.

Dr. Day describes how synesthesia websites and Facebook groups have been popping up around the world, too—in Belgium, the Netherlands, and Sweden, for example. And Wikipedia now translates the synesthesia page into Czech, German, Greek, Spanish, French, Italian, Hebrew, Dutch, Hungarian (Magyar), and Japanese. I'm online friends with at least a dozen people from other countries—from Germany, Belgium, the Netherlands, and Russia—due to synesthesia groups on Facebook. The camaraderie is wonderful, and the sense of unity, despite the language gaps and cultural differences, remarkable. We are not bound by nationhood or ethnicity or ideology, but rather the biological gift we have in common, even across these many thousands of miles. There is a shorthand among us.

I ask several of these new friends around the world for their views of how the Internet has affected their experience with synesthesia. Regina Pautzke, president of the German synesthesia association, says the web has changed her experience markedly. "The effect of the Internet on connecting people and sharing experiences is huge. In conversation with others you might be pointed to an aspect of synesthesia you did not see before. I myself found out about more types of synesthesia I have and did not realize before," she says. But more than that, she says, the formation of these forums and groups has helped improve both the quality of information that gets put out there and the way it's handled in the media. "What I really like about the Internet is that by providing serious information on the topic, we are able to influence the media and the quality of reports on synesthesia." Her online forum collects all kinds of information, from researchers and as well as synesthetes themselves, and it has been invaluable to journalists and students asking for help.

Dr. Alexandra Dittmar, the German editor of *Synesthesia: A Golden Thread Through Life*, in which she presented twenty-one real-life accounts of how synesthesia affects people's orientation, points to the diversity of media that now exists online about synesthesia. Her lovely visage, complete with a butterfly that seems to have alighted on the frame of her eyeglasses, greets me each time I log onto Facebook. "In my opinion," she writes, "the Internet provides a richness in opportunities to get information about synesthesia in various forms (written, pictures, videos, etc.)." Clearly this is a much richer and wider landscape than existed just ten years ago. She also likes the ability to meet other synesthetes, "just online or maybe later even personally. It also provides many opportunities to find institutions and scientists doing research on synesthesia, and it makes it much easier to find events, lectures, exhibitions, etc., dealing with synesthesia."

Dr. Dittmar says the advent of Web 2.0 and social media had an even more profound effect on the way synesthetes interact worldwide, both with each other and with the public. "In its beginnings, the Internet was mainly an exchange of information. Interaction took place mainly through emails (mailing lists), news groups, and the like." Today, she says, its focus has drifted toward more overt interaction, "like forums, blogs, and Facebook. And Internet users are more able to create their 'own' Internet. They can, for example, personalize applications according to their interests and needs, or they can more easily post and share videos and pictures on public websites." This has also helped synesthetes shape their culture, she says. "It began with mere information on synesthesia, and communication was possible through mailings (mailing lists or newsgroups) which did not directly show up on websites. Today, we are able to create online profiles as synesthetes, sharing our self-made videos, pictures, texts, and the like, in and for the public."

Spanish synesthesia researcher Mj De Cordoba wrote me from a computer station in Granada. I notice that her words in Spanish to me are in color, as well. She believes that the Internet is a great investigative tool (*herramiento*, Spanish for "tool," is a rich burgundy to me, whereas the English equivalent is a cool yellow due to the different spelling), although one must always have a critical eye and compare information. "To be able to put people in contact with each other from all parts of the world who are interested in the same things as you, is fortunate. It's possible to share experiences and furthermore, to see that we are not the only ones who have these types of experiences, which is a relief. It's also enriching. For me, to find the collaborations of Dr. Sean Day and Dr. Ed Hubbard in 1999 was fortuitous" (my translation).

James Wannerton, president of the UK Synaesthesia Association in England, got an early start with the useful communications medium. He writes to me electronically from across the pond: "As an enthusiastic early user of the Internet back in 1991, in the days when the only services offered were email and Usenet access, I have observed and experienced the medium grow into the indispensable and remarkable tool it is today." A man with profound synesthesia, including gustatory-lexical synesthesia (in which one tastes words), he says he made contact with his very first fellow synesthete via one of the early Internet bulletin boards on the topic. The event was so meaningful to him that he remembers the date, "November 17, 1991, to be precise—an event that had an immediate and profound effect on my future thinking and direction. By virtue of this alone, the Internet really does have a special place within my being."

He credits the Internet as the reason behind the recent explosion of interest in synesthesia, "a process begun 'manually' via the printed word by the likes of Professor Simon Baron-Cohen and Richard Cytowic in the early 1980s," he points out. "The Internet's ability to disseminate information easily and quickly across geographical boundaries served to accelerate this process at an incredible rate. I well remember using one of the very first search engines to seek out the term 'synaesthesia' and the return was a total of six hits. Today, entering 'I can taste words' or 'hear in colour' into a search engine produces a vast array of references, some misleading, it has to be said, but the vast majority allow the 'Googler' an enlightened view into the fascinating world of the synaesthete."

Perhaps the most rewarding aspect of the Internet for Mr. Wannerton is its ability to help people who are just coming to terms with their gift. "Myself and Dr. Sean Day of the American Synesthesia Association have received hundreds upon hundreds of [online] contacts from synaesthetes who are truly invigorated to learn that 'this thing they do' has a name, and that there are others who perceive things in a similar way. This information and access to the world's top experts on the subject can and has been life-changing for some, and it is this that I personally find the most rewarding aspect of my own work via the Internet here in the UK."

Researchers are also reaping the rewards with what Mr. Wannerton calls "a rich seam of subject matter." He describes how neuroscientist Dr. David Eagleman has amassed a "database of synaesthetes [the Synesthesia Battery], which provides an invaluable service in the study of large-scale analysis as used alongside ongoing neurological and genetic research." He adds that the web has made it possible to forge strong and lasting relationships around the globe. "This has led to the international exchange of experiences and ideas between researchers with a view to initiating new avenues of research and to lay foundations for future cooperation. Another very important function of the Internet is that it fuels interest in the study of synaesthesia, providing us with new researchers with new ideas. Yet another is that it allows for all opinions and ideas to be aired—good and bad. If these opinions and ideas are deemed wrong, the Internet provides the means to put across an alternative idea—never a bad thing."

Mr. Wannerton says that the Internet will become increasingly useful to those with synesthesia. "The Internet can and will bring remote and disparate communities together via increased use of remote access, podcasts, and the like. The amount of interest generated by social networking sites such as Facebook is phenomenal, an offshoot of which is providing virtually instant responses to research requests. There are already synaesthesia-themed interactive online games out there, and who knows, maybe one day we will unearth a new kind of Internet-induced variant of the

condition such as 'Mario-to-taste' synaesthesia." Mr. Wannerton maintains a witty and robust website on the topic:
jameswannerton.com.

Russian synesthesia researcher Dr. Anton Sidoroff-Dorso
spoke with me through Facebook. He thinks that the 'net
not only helps identify all the known types of synesthesia for
"experiencers," but will actually unveil new forms of synesthesia, similar to James Wannerton's humorous "Mario-to-taste"
variety. "As a researcher I would say that with the advent of
the [web] and with the new younger and younger digitalized
generation coming along, we could anticipate new types of
synesthesia. Virtual reality, being materially ungrounded, is
likely to produce novel challenges for human minds in terms
of time and space perception and attribution. Anthropologi-
cally speaking, the Gutenberg Galaxy [the world of print] has
brought about grapheme-to-X synesthesia while these days,
perhaps, new Internet-based practices will crystallize into
new inductive categories such as domain extension-to-X,
or avatar-to-X, or even site-to-X synesthetic experiences.
Likewise countries or names: one might soon have mauve
dot-com or grass green dot-ru."

In addition to friendships and potential new synesthetic
experiences based on brand-new stimuli, the Internet has
also sped up the research currently under way that will
aid the many people around the world who are in search
of more answers about their unique trait. The Internet
makes things happen larger and faster than ever before.
"Contrary to how things were twenty years ago, when

synesthesia began to reemerge as a topic of research, [and] when researchers would have to travel to a university library and spend hours to only track down a few books and articles toward answering [their] questions," Dr. Day explains, "now they can get on the Internet, acquire handfuls of names of major synesthesia researchers in a matter of minutes, and email off questions or otherwise contact the experts for direct correspondence." He says this has produced an amazing explosion in the last decade, several orders of magnitude greater than what has ever been seen previously. It has sped the research process up remarkably.

It has also manifested changes in real life, he says. "Through these organizations and website resources, with the help of email, smaller, regional synesthesia groups have emerged in such a way as to become 'non-virtual' but far more 'concrete.'" For example, he says, in the San Francisco Bay area, "synesthetes have been able to contact each other and organize group meetings and events, such that there is an emergent (albeit small) San Francisco synesthesia community. Similar things have happened in New York City, London, and the Minneapolis/St. Paul area, and [are currently happening] in Los Angeles, Boston, and the Seattle area." Most but not all of the groups have been facilitated by university-affiliated researchers helping to put local research subjects in touch with each other when they ask for contacts, he says.

Both Sir Robert and Dr. Day are concerned that there is a Western and economic bias to attaining information via the Internet due to the predominance of English and the lack of access to computers in the far reaches of the globe. In addition, more is known about synesthetes in the West, though it is likely a global trait. Hopefully, as more people gain access the tool that Sir Robert developed, and which Dr. Day and other researchers and leaders are using to provide a public service so effectively, those disparities will diminish. And though a virtual community is great to have when one has such a unique gift, Sir Robert has said the Internet will never replace face-to-face human interaction. As he said at the second World Wide Web Conference in Chicago, "There is no such thing as a virtual beer."

—

Talking with Sir Robert and other synesthetes using the Internet makes synesthesia seem so accessible and understandable, and yet there is still something about it that is elusive, intangible. In fact, speaking of it so much and reducing it only to a function of language, when it is so much more than that, is beginning to worry me. Perhaps it's because it was this precious secret I was able to keep, just for myself, for so long. I wonder if it could disappear under this looking glass if I'm not careful with it, as often happens with recurring dreams when you finally speak of them. I really don't want that to happen. . . .

4
—

EMERALD BAYS AND BLUE NOTES

The rolling opening piano melody of "Summer, Highland Falls"
begins, and it's as though a drop of pink paint has fallen into an
aquarium of water in my mind's eye, it is pulled to and fro by an
invisible string, never completely dissolving into the surrounding
liquid. As I meditate on this, my favorite Billy Joel song, I am
delighted by the accompanying synesthetic impressions. Various
form constants (those shapes of photisms synesthetes see) open
their peacock tail feathers for me—pointy white things that are

assembled all in rows, like the bristles of a flat man's hairbrush; a
flat yellow "tape" that comes down from above in the right upper
quadrant of my interior tableau and then flattens out horizontally.
Most remarkably, as Mr. Joel sings the word euphoria, a flame ig-
nites in the center of this visual field, impossibly black in the center
and licking about on the edges in magenta and gold. I wonder what
color this song is to its synesthete composer. . . .

There is a longstanding rumor in the synesthesia community that the great singer-songwriter Billy Joel shares this
gift. I realize that his testimony would mean a great deal
to my newfound community of online and real-life synesthetic friends, and so I decide, on this next leg of my
journey, to see if the Piano Man himself would be willing
to offer some insights. Since I've embarked on this journey, however, I've harbored a secret fear that speaking so
openly and analytically about synesthesia—taking it apart,
examining it, putting it back together—will take some of
the mystery out of it. So I am heartened and encouraged
when he tells me that that he wants to be careful not to get
too clinical about his rare color-to-music gift, lest it vanish.
And who would blame him? Synesthesia has helped him
become the prolific songwriter the world loves. His interview means all the more to me because I know this fear,
and I think most synesthetes would, too.

Talking with me from Miami between tour dates with
Sir Elton John, Billy Joel publicly admits his unique gift for
the first time. He tells me he has explored his synesthesia

"a little"—but purposefully not enough to demystify it. "If I figure it out, somehow the sorcery disappears," he says. "I don't want to become formulaic with it; I kind of like the spontaneity and the mystery of it all. It's very intriguing to me." In saying this, Mr. Joel is expressing something that all of us synesthetes feel—namely, that there is an ineffable quality to the gift. To ascribe it to "unpruned neurons" or "a lack of chemical inhibition"—two of the dominant theories of its cause—somehow deflates the mystery. For me, this mystery resides in consciousness, not just anatomy.

Dr. Richard Cytowic, a pioneering neuroscientist, believes that synesthesia is above all an emotional experience "accompanied by a sense of certitude (the 'this is it' feeling)," which he compares to William James's description of religious ecstasy, and to noesis, an illumination accompanied by a feeling of certitude. Although brain scans have shown synesthesia to be active in the *fusiform gyrus* and *angular gyrus* (discrete areas of the brain, behind our ears and above them, respectively), he has theorized that there is another link between synesthesia and the limbic brain, which, oddly enough, is associated with emotion and a feeling of lucidity. It may "live," then, in a place that is at work during sleep and deep meditation.

Interestingly, though he himself doesn't meditate, Mr. Joel says the peer he feels the most affinity with in regards to the creative process is Sting, who uses trance-like states for inspiration. "Sting has expressed something similar to what I think happens. [H]e's able to tap into a different consciousness. He studies yoga so he's able to meditate with a lot more discipline than most people. My feeling is that [synesthesia] exists in a different plane and we tap into it somehow. . . . I think I do it in a dream state."

When the Rock and Roll Hall-of-Famer pauses to create, he has exceptional experiences. He sails the full spectrum of colors on his creative sojourns: His ballads are born in coves of azure and emerald, but his rock music is forged in fiery landscapes of the "hotter" colors. Amazingly, he often dreams the hits that have become so much a part of the soundtrack of our lives; the notes appear against a field of amorphous and abstract colored shapes in his sleep. In his waking world, he associates musical genres with specific colors, too. "I would say the softer, more intimate songs—there's 'Lullaby (Goodnight My Angel)' [written for his daughter, Alexa Ray], 'And So It Goes,' 'Vienna,' and another song called 'Summer, Highland Falls'—when I think of different types of melodies which are slower or softer, I think in terms of blues or greens." Conversely, songs with a heavier beat and a faster rhythm (think: "It's Still Rock 'n' Roll to Me" and "We Didn't Start the Fire") suggest the red-orange-yellow end of the spectrum, Mr. Joel explains. "When I [see] a particularly vivid color, it's usually a strong melodic, strong rhythmic pattern that emerges at the same time. When I think of [these] songs, I think of vivid reds, oranges, or golds."

Though peaceful aqua hues for ballads and brash scarlet tones for rock music seem logical even to non-synesthetes, I know he's not being metaphorical here. Few synesthetes have the same colors for musical genres, notes, letters, or numbers; he might just as easily have said the opposite, or used purple for both. It's completely unique to each person. There is no Rosetta Stone for the infinite number of possible pairings and sensorial correspondences experienced by synesthetes around the world. To illustrate this, I remember an amusing dinner conversation with a charming synesthetic artist, Carol Steen, who first mentioned Mr. Joel as a possible synesthete after hearing it from his former neighbor, noted photographer Marcia Smilack, who is herself a synesthete.

"What color is your A?" Carol asked me one night in Chelsea.

"Yellow," I replied.

"My dear, A is definitely pink. Perhaps there are vitamins you could take."

We synesthetes are very attached to our individual color associations; after all, they've been with us throughout our lives. We often speak of the pairings with great affection, and even rise to tones of reverence when speaking of a favorite color among them. Though Mr. Joel thinks that every hue he sees for his music is gorgeous, he says that the rich greens resonate most for him. "I think I have an attraction to different colors at different times," he says. "The one that's most attractive to me, more often than not, is dark green for some reason. I have an attraction to very dark green: hunter green, kelly green, royal green, deep green." He says the words as if he's tasting the notes of a fine wine.

Coincidentally, Marcia had a conversation with him years ago that led her to believe that he is a member of the "tribe." Marcia is a "reflectionist" who takes stunning photographs of reflections on water. She recalls: "We were looking at one of my photographs and I explained how the green section on the left under the horizon sounded like chords, while the two streaks of pink lightning above on the right sounded like grace notes. He said he knew what I meant and I could tell he really did, so I felt he was also synesthetic."

I decide to ask Dr. Larry Marks of Yale about Billy Joel's sound-to-color synesthesia. Dr. Marks has been studying synesthesia for many years. At the time we spoke, he was the fellow and emeritus director of the John B. Pierce Laboratory, and professor of epidemiology and public health as well as psychology at the Yale University School of Medicine. His laboratory, appropriately enough, is devoted to sensory information processing. I like to refer to the very paternal and affable Dr. Marks as a synesthesia ETB (Early True Believer), to use Internet parlance. I ask him about Mr. Joel's particular form of synesthesia. "Some synesthetes 'see' colors when they imagine sounds, whereas others have to hear the sounds to see the colors," Dr. Marks explains. "There is considerable interest in synesthesia research regarding the sensory, cognitive, and imaginative factors involved in synesthesia." I can tell that Billy Joel's synesthesia has piqued the interest of the seasoned clinician because it is often present in his inspired, creative state.

Those gentle blues and greens also figure into the softer parts of language for Mr. Joel. In addition to his sound-to-color synesthesia, he also has the grapheme form (graphemes are letters, numbers, or symbols). "Certain lyrics in some songs I've written, I have to follow a vowel color. A strong vowel ending, like an A or an E or an I [which he uses at the end of a line when he wants to hold a note—a device he says he learned from The Rolling Stones' Keith Richards], I associate with a very blue or very vivid green. I think reds I associate more with consonants, a T or a P or an S. . . . [It's] a harder sound. These [letters] are what I associate with reds and oranges."

Interestingly, vowel sounds themselves, "sometimes the letters of vowels, are often strong inducers of synesthesia," observes Dr. Marks. Indeed, French symbolist poet Arthur Rimbaud composed a stunning poem about colored vowels titled "Voyelles." Along with the works of fellow symbolist Charles Baudelaire, Rimbaud's work helped to popularize literary synesthesia.

The two symbolists were not natural synesthetes, however; they dabbled in drugs to achieve the experience. A portion of that poem is reproduced here:

"Voyelles"
A black, E white, I red, U green, O blue: vowels,
I will one day declare your latent births;
A, black velvet corset of exploding flies

Which buzz around cruel stenches,

Gulfs of shadow, E, candors of vapors and tents,

Lances of proud glaciers, white kings, shivering umbels. . . .

(From The Cambridge Introduction to French Poetry, translation
by Mary Lewis Shaw)

Dr. Cytowic, who, like Dr. Marks, ushered in a new era of research into the trait (with his groundbreaking 1989 book *Synesthesia: A Union of the Senses*, the first-ever English offering on synesthesia, as well as *The Man Who Tasted Shapes* in 1993), wrote on letter-color priorities in synesthetes and non-synesthetes in his recent release with Dr. David Eagleman, *Wednesday is Indigo Blue: Discovering The Brain of Synesthesia* (MIT Press, 2009). In this book he suggests that sensitivity to pronunciation, as Mr. Joel describes, is an even rarer form of what is generally known as grapheme-to-color synesthesia: "Sensitivity to pronunciation seems to hold for only about 25 percent of grapheme-to-color synesthetes polled (by researchers Simner, Glover, and Mowat in 2006) and implicates an auditory component to their synesthesia," he writes. "In other words, some grapheme-to-color synesthetes have an interaction with auditory parts of the brain, whereas for most their synesthesia is based only on the written letters."

Mr. Joel first experienced these auditory phenomena as a child. He soon realized they were unique, he says. "So when kids would come in to school and say 'I had this dream about a monster,' or 'I had this dream that I saw you or somebody died,' [or] 'I had a nightmare,' I thought to myself, *Gee,*

I have a completely different kind of dream. I dreamt a melody or I dreamt a great rhythm or a chord pattern or symphonic fragment for a song. It was always music but it wasn't always the same kind of music. I have had literal dreams like [those of] other people but more often than not it's an abstract kind of dream." Part of this abstraction is dreaming in shapes, he explains. "They're amorphous shapes, not often geometric, and [they] don't seem to have a logical pattern to them. They could be proto-plasmic shapes, amoebic shapes; there [are] a lot of curves, there [are] jagged edges sometimes . . . No particular well defined shape to [any of them]."

I've experienced this a number of times myself, but in a waking state and in a passive, decidedly uninspired way. When I listen to Fiona Apple's song "Sullen Girl," for example, I see a variety of jagged shapes morphing about in silvery and pearlescent-white shades. In their buoyancy they seem carried by a wave of invisible water, and not just because of the singer's highly synesthetic lyrics. They "pour" from somewhere. I get similar moving imagery when listening to Leonard Cohen's "Hallelujah," but the forms are mostly amorphous undulations of yellow and gold. If you are not a synesthete and you wish to experience this yourself without using drugs, watch the scene in the film *The Soloist,* in which *Los Angeles Times* reporter Steve Lopez takes homeless musical prodigy Nathaniel Anthony Ayers to a rehearsal of the Los Angeles Philharmonic. Mr. Ayers closes his eyes in ecstatic reverie while Dreamworks Animation in cooperation with Paramount Studios supplies the synesthetic animated magic of what he sees in his mind's eye. Although Mr. Lopez does not believe that Mr. Ayers, a schizophrenic, really is a synesthete (the scene is not the same in his bestselling book), the filmmaking trope is beautiful and realistic nonetheless. It also demonstrates the joy that Mr. Ayers, once a promising Juilliard student, felt listening to live orchestral music once again after years on the streets.

Mr. Joel also mined his abstract dreams on his 1993 album *River of Dreams*. The song "In the Middle of the Night" uses a water-themed description to explain one dream in particular. The songwriter says he's still not sure to this day what that song means; it was just a dream that emerged, fully formed, from his mind. He didn't actually want to write the song but he couldn't shake it off, it haunted him so. "When you wake up singing a song and having a certain rhythm running through your mind and you can't get rid of it, it means something. I've always found it's best to follow up [on] the initial impulse when something is that strong," he says.

Lauren Lawrence, the "Your Dreams" columnist for the *New York Daily News* and author of *Private Dreams of Public People* and several other dream books, believes that Billy Joel's dreams are at his creative heart. "In that dreams are pictures, it should come as no surprise that there are different dream styles: Some dream landscapes have the distinct brushstrokes of realism and are easily identifiable, while others are more suggestive, veiled in metaphor, and thus surrealistic in nature," she explains. "There are impressionistic dream images that are imprecise and therefore more open to suggestion, and there are those that are abstract, involved with the shape of things to come. Recognizing the many painterly styles of dreams is of interpretive value as it is self-referential of the dreamer." Similarly, she says,

> the tones or colors of a dream, as described by a dreamer, serve to delineate emotional content. So when Billy Joel dreams of reds and oranges—color manifestations of anger and aggressive impulses—it is not unusual. What is unusual is the auditory component: He interprets the colors musically, and hears rock melodies. By listening to the colors of his dream Mr. Joel has found a way to utilize the primitive emotion of anger to inspire creativity. It is this synesthete's way to tame rage through song . . . [F]rom a psychoanalytic perspective, the "brightness" or "darkness" of colors that Joel associates with the tonality of sounds relates to his mood. It is as if he has codified the day residue (the antecedent stimulus of a dream) in terms of color, shape and sound as opposed to feeling. Similarly, dreaming of shapes is the synesthete's way of conceptualizing form as opposed to content. It is a way of profiling—of cutting to the chase. And a way of seeing the forest and not the trees. It represents a particular mindset which concerns itself with the whole and not the sum of its parts. The "amorphous," ill-defined designs of Billy Joel's dreams indicate the wish for transformation, the wish to shape something formless from inception, and this process is at the heart of creativity.

To Dr. Cytowic, it is significant that Joel's dreams have a residual value. "[This is] an example of synesthesia influencing his art, in much the way [that] Carol Steen paints or sculpts the photisms she sees," Dr. Cytowic remarks. (Photisms are the tiny points of colored light that synesthetes see.) Carol Steen is a New York City–based painter and sculptor and cofounder of the American Synesthesia Association. She has five forms of synesthesia—grapheme-to-color, sound-to-color, smell-to-color, touch-to-color, and pain-to-color—and paints or sculpts her visions according to how visually interesting they are. "My two favorite forms of synesthesia, the ones I usually use to create my paintings and sculpture, are colored sounds and colored touch," she says. "When I listen to music I prefer it to be loud and, if possible, to be heard live, because to paint what I can remember from a concert is tricky for me; I always forget much more than I [would like]." Even when she can't get to a live performance, Ms. Steen has found a substitute. "I can play a piece of music on my excellent speakers at a loud enough volume to make the heard colors bright." Though she still prefers live music and says it has "the best colors," the "richest textures," and the "most subtleties," listening to recordings has the added benefit of letting her replay sections of music she may have forgotten.

Like Ms. Steen, Mr. Joel sometimes has difficulty retrieving his inspiration immediately, but says that the colors seem to act as a mnemonic device, just as they do for Sir Robert. Synesthetes are known to have superior memories precisely for this reason. As a child, if I was facing an exam and couldn't remember that the United States entered

World War I in 1914, I would just envision black-brown-black-red and work backward from there. In adulthood I find it works well with phone numbers and names, too. Billy Joel believes that the ease with which he can access his musical dreams depends on the quality of the colors involved.

"I think it has something to do with the brightness or the darkness of the color associating it to the brightness or darkness of the tone or sound," Mr. Joel says. And then, there's always technology. "I've slept with a small cassette tape recorder next to my bed for years. Sometimes I [. . .] wake up in the middle of the night, to capture this on tape but it never makes any sense the next day. The next day it sounds like gibberish. There's no context to it. Essentially I'll end up singing into the tape recorder, I'll wake up [and] say, *Gee, I want to remember this,* and I'll sing a melodic fragment into the tape recorder, not remembering the next morning what was the rhythm, or what was the chord progression, what was the context of the melody, how would it start, how would it end. So it's just fragments. . . ." Dr. Cytowic finds this fascinating. "Like dreams we write down in a dream diary in the middle of the night, they often don't make sense to us in daylight. Or we [miss] some crucial, emotional elements of what was consequential about the dream."

Mr. Joel says that songs will sometimes reappear the next day, but sometimes they won't return for months or even years. "The song 'Just the Way You Are' is a song I had dreamt and forgotten. And I was having a business meeting just a few years after I had this dream. And right in the middle of this very boring business meeting—I think it was an accountants and lawyers meeting—my mind was wandering a little. All of a sudden I remembered this song; well, I didn't know that I was remembering it. All of a sudden this song popped into my head, and I told the people at the meeting, I said, 'Look, I've gotta go home right now—I have an idea for a song!' And they said, 'Go, go, go, write the song!'" The storyteller/bard goes on to explain that he wrote the song fairly quickly and then asked himself, How the hell did that happen? On reflection, he realized he had heard it before. Then the "aha moment": Yes, it was a synesthetic vision in a dream! The blues and greens and their various amorphous shapes had encoded it in his mind so that he could retrieve it later. Observes Dr. Marks, "Color seems to play an important role in Mr. Joel's musical memory, and to be sure, in the emotional nexus of his music."

Dr. Cytowic has written that synesthesia may be a conscious manifestation of what is a normal, holistic cross-sensory joining normally occurring automatically and unconsciously in other people. This seems to suggest a more porous interface between synesthetes' conscious and subconscious minds than exists in other people. This is a fortunate happenstance for Mr. Joel, whose inspirations don't necessarily come during "working hours," and who must recall the inspirations that sometimes come from beyond his waking mind. "It's not the kind of job where you can just shut the door and leave the office at 5 o'clock," he explains. "It's a 24/7 process and your mind just doesn't stop when you're trying to work through something that's problematic that you may not get. . . . [I]t continues subconsciously. Sometimes I'll dream an entire symphony from beginning to end. It's as if it was all composed in one fell swoop [. . .] in the dream," says Mr. Joel.

Remarks Dr. Cytowic, "This is very typical of creative works that appear in dreams—they are clever, wholly complete, and emotionally satisfying." Mr. Joel theorizes that when one is working and conscious, everything is very tidy, organized, and logical, "[but when] you close the office and you leave for the day, all the little elves that have been hiding in the office come out and start playing with your equipment—they get on your computer, they get on your typewriter, they get on your piano, and they just take over. And it's not a logical process. It's like a bunch of kids ransacking a tool shop." In an unconscious state, Mr. Joel says, all bets are off; it's delightful anarchy. He finds this to be somewhat helpful sometimes in that he's able to take the blinders and the restrictions off: No rules apply and anything goes.

He adds, "And it's a much freer state, [this] subconscious or [. . .] unconscious state. That's how it seems to work for me. I have written songs [when I wasn't dreaming], although I suspect that the germ of the song comes from a dream that I have recognized or that I remembered or recalled." Sometimes he'll have written a song, an inspired song, and he really wonders where it came from. "There's almost a sense that I lifted my head up into this rarefied stratosphere and the idea came like a Promethean moment. I don't always know how it happened but I know that it definitely happens from a dream state. And it's just an assumption I have because I really haven't analyzed it all that deeply. But when I am in a conscious state and I do write and I do come up with something that it may have germinated in a dream state."

Mr. Joel says he is grateful that his musical parents paid for piano lessons for him at an early age because it enabled him to realize his musical dreams, literally. Music clearly runs in the family: His father was a pianist, his mother was musical, and his half brother is a conductor based in Vienna, Austria. "It was a secondary language in my house, in my home life, in my household." And it seems that he has passed both of his gifts onto his talented daughter, Alexa, the little girl who sang all those songs to him, sailing on an emerald bay in his lullaby dedication. Synesthesia runs in families and is often passed from father to child. Mr. Joel concurs: "My daughter has said she has had dreams of music. . . . We've shared stories of, 'You know, I woke up and I had had this incredible dream. I dreamt a symphony and now I don't remember what it is and it's very frustrating.' It drives you nuts!" When he dreams, say, a symphony, Mr. Joel says he knows that it's not a literal or linear thing, like being in a concert hall or listening to a recording. Rather, he states, "I'm dreaming of a color being present during the music being played or the music being heard. . . ." Our dream expert, Ms. Lawrence, solves the mystery: "He has used color to encode memory. So what he hears is of emotional value and similar to what one visualizes in a dream."

Mr. Joel wonders if his gift would have flowered had he not had piano training from the age of four. "I made the connection immediately with a form of communication, with a form of almost talking to myself." It opened up a rich inner life, he explains. "When I was a little kid I would just walk over and start banging on it. If I wanted to write a storm song I'd start playing—just banging on the low notes for the thunder and banging on the top notes for the lightning—and after four years my mother said, 'That's enough of that song. Let's go to the piano teacher so you can learn how to play this thing right.' All I know from my own experience is that emotions [and] [. . .] logic [are] synthesized somehow into music in my mind. I can feel as if I had a [real] experience through hearing music. Even when I'm consciously listening to music, listening to composers like Beethoven or Rachmaninoff or Debussy, I go through an emotional experience as if I've had an interaction with another person, as if I've just had a love affair or [. . .] a great sadness or a great joy. It's the same exact feeling. I can go into an ecstasy [. . .] [or] a depression just from the way my mind responds to music, my emotions respond to it." Mr. Joel says he is so sensitive to operatic music he often finds he doesn't need a libretto to know what's going on. "I kind of comprehend what's going on [based on the music alone]."

In the early 2000s, Mr. Joel returned to what he believes is a purer form of his synesthetic expression: classical music. "There is an album of piano pieces [*Fantasies and Delusions*] that I composed. It came out not long after 9/11. These were all piano pieces that I was compelled to publish, and I call it *Fantasies and Delusions* because it's a little audacious for someone like myself who's worked in rock 'n' roll and pop music to all of a sudden present 19th-century style romantic piano pieces as if I were a contemporary classical composer. But it's something that's always been a desire of mine to do. And these are pretty much completely dreamt-through pieces. I didn't sit down and fuss a lot with form and structure and development and variation consciously. They were pretty much presented as they were dreamt, which is another reason I [gave the collection that title]." He says these compositions are the purest translation of synesthesia he has produced because lyrics don't appear to him in dreams or inspired states—just melodies carried on colored forms. Though he writes lyrics later in an attempt to convey the emotion of the inspiration, he feels the original synesthetic impression is ineffable, so no words, no matter how poetic, could ever quite capture it.

"In the piano pieces in *Fantasies and Delusions*—they're all subtitled, by the way—there is a connection to a human or a personal experience that this music is describing. One of the pieces is called 'Soliloquy on a Separation' and [it] is supposed to emotionally describe the end of a visitation with my daughter when I was getting divorced. And it might help to explain what the piece is trying to express because I realized when I [. . .] dreamt these pieces that they all came from an actual experience and I was trying to musically work it out. . . . I'm by myself trying to translate my own work." In another beautiful piece in the collection, titled "Star-Crossed Suite," a love affair is recounted in three different movements, he says. The first movement is all about desire and longing, and is titled, appropriately enough, "Inamorato." The second movement is the physical consummation of that desire, and is titled "Sorbetto." And the third movement, titled "Delusion," has to do with that time when the rose-colored glasses come off and people see relationships for what they really are. Each of these pieces was dreamt, he explains.

He says as he matures he's willing to consider the possibility that something beyond this present reality exists. "At one time in my life I rejected all things spiritual and I thought I had an answer for everything. But now at this point in my life—I'm almost sixty—I think there are different spiritual planes that I can't explain, that I don't understand, and that I've never seen an explanation for, and I'm willing to accept the possibility [that these things may exist]."

—

With testimony from the man responsible for so much of the soundtrack of our lives now secured, I apply to a prestigious new writers colony to work on a related essay. My task during this leg of my journey was to "keep the sorcery" of my gift intact. Now I must ask myself if I can make the transition from newspaper journalist to author. I know that I will have to reach deeper into myself and reveal more of my own impressions in order to do so. . . .

5

—

A BLONDER SHADE OF SYNESTHESIA

I pack my bags for the Norman Mailer Writers Colony in Provincetown and find myself reflecting on the late Mr. Mailer's words: "Writing is spooky. There is no routine of an office to keep you going, only the blank page each morning, and you never know where your words are coming from, those divine words." I wish he were still alive to talk about synesthesia, for all his macho swagger, Mr. Mailer clearly had a deep, seeking side. Synesthesia is a spooky topic if ever there was one. Where are the words coming from, indeed? And why do mine appear in full color? What I don't realize is that through his colleagues at the Colony, Mr. Mailer will have some answers for me, even in death.

As I stand in the living room of Norman Mailer's Provincetown, Massachusetts, home, Dr. John Michael Lennon reaches for the bookshelf and hands me a paperback version of one of Mailer's books. The beautifully decorated room is now also filled with a conference table and chairs for our class of six people to study New Journalism. To me, it's tingling with possibility. I am fortunate to be here after being chosen from hundreds of applicants by the Michener Center at the University of Texas and the Colony for a week's training. Because my application contained an excerpt from this book, Dr. Lennon knows that this is my passion. My professor believes that his friend Mr. Mailer certainly used synesthetic metaphors in his writing.

Dr. Lennon is a Mailer scholar and intimate who wrote *Norman Mailer: A Double Life.* The bearded, scholarly man, who is emeritus vice president and emeritus professor of English at Wilkes University in Pennsylvania, seems melancholy about the loss of the late Mr. Mailer, which I find touching. The Colony is the way he and Mr. Mailer's other close associates and survivors keep his memory alive, and it is a fitting tribute to a man who was known for helping aspiring writers (he once even gave a nurse some pointers when he was in the hospital).

I'm clutching the beautiful orange volume of *Ancient Evenings* in my hands now, sitting at the table. Dr. Lennon opens the book and points out a passage. In it, the main character, General Menenhetet Two, is adjusting to new and more expansive sensory perceptions as his soul is released upon death. No longer corralled by and channeled through the sensory organs, his perceptions join together synesthetically as he tries to make his way in this new reality. "I felt larger, as if my senses now lived in a larger space," the general remarks. Even though I'm not in the Land of the Dead, I recognize that feeling immediately. This is what it feels like to be a synesthete. It's similar to the way your sense of personal space can expand when you jump into your car— suddenly you are you, but also the entire automobile. I feel the joy of someone understood by her teacher and, further, enriched by him. With his adroit use of metaphor, Mr. Mailer's imaginings soar to a cross-sensory world greater than the sum of the individual senses. And therein lies his genius. As Aristotle said in *Poetics*, "The greatest thing by far is to be a master of metaphor." It is "a sign of genius, since a good metaphor implies an intuitive perception of the similarity in dissimilar."

The New Journalist masterfully evokes the confusion of this newly released soul, giving the general a kind of "nasal dizziness" in the following extract: "Merely figure the vertigos of my nose when the empty cavity of my body (so much emptier than the belly of a woman who has just given birth) was now washed, soothed and stimulated, cleansed, peppered, herbified, and left with the resonance through which no hint of the body's corruption could breathe." As well, the plants and spices used in the embalming ritual are heard crying out, singing in this psychedelic mezzanine on the way to the afterlife: "The myrrh even made its clarion call. A powerful aromatic (as powerful in the kingdom of herbs as the Pharaoh's voice) was the myrrh laid into the open shell of my body. . . . Like rare powders added to the sweetmeats in the stuffing of a pigeon, were these bewildering atmospheres they laid into me. . . . Oh, sweet smelling soul of the Great God." Can a soul smell sweet? Can plants and spices cry out? If one is a synesthete, certainly.

—

After class I walk Commercial Street to the downtown area, passing beautiful gardens, art galleries, and a modern Internet café that I will make the end point of my early morning constitutionals each day. I'm still a frustrated "yeti," a displaced member of the young, entrepreneurial, tech-based generation, and still mourning my former job as a web editor until the bubble burst in 2000. But this colony experience is a tribute to the immortality of print and the value of books and of human contact. And it is beautiful here and soul-quenching, a far cry from my less literary life in breaking news.

There is a sacred geometry to Provincetown. The Masons knew it when they designed the layout. And Stephen Borkowski, chair of the Provincetown Art Commission, alludes to it when he notes that there aren't too many places on Earth where the land curls back on itself. He has come to visit us so that we may learn the spirit of the town through two of its most prominent citizens and then write about it as an exercise. "P-town is as the nautilus shell," I type later on my laptop, alluding to the Fibonacci sequence and golden ratio of its structure found in all of nature and even in musical scales. And the center spiral seems to be this table—both real and symbolic—that we are sitting at on the bay-front terrace of Norman Mailer's former residence in the center of the harbor's curl. Emmy Award-winning director, author, and photojournalist Lawrence Schiller, president and cofounder of the Norman Mailer Center and Colony, is filming us for posterity today and every day we're here. He's omnipresent, even when we walk down the street outside of the estate. He even shoots over my shoulder as I write in my notebook.

Provincetown is intellectually serious, but it also has a carnival aspect. As if to illustrate this, Mailer, the great thinker and commentator of his generation, loved to walk a tightrope near this table, his solid torso providing a reliable center of gravity to traverse the span. His good humor was appreciated by many in the town, including arts figures Stephen Borkowski and Chris Busa, whom I will later meet in order to better understand the spirit of the town as well as the great author. Mr. Borkowski says that a date dragged him "kicking and screaming" to Provincetown for the first time in the 1970s; he returned with a more serious partner a decade later and decided to stay. Initially he found the atmosphere discordant, "a little too frantic, a little too scene, too structured," he says around a table on Mr. Mailer's storied terrace, looking out into the harbor, the morning sun making his blue eyes sparkle. Later, it was "let's spend a week in Provincetown, or our weekends, or a season."

Beyond the layers of the social scene, the retail scene, the tourist scene, he started meeting some interesting people, such as the man who could whip up risotto for twenty at a moment's notice. He began to make friends. This once-conservative grandson of Polish immigrants gets dishy and talks about the time Marlon Brando fixed the toilet in Tennessee Williams's place so that he'd cast him in *A Streetcar Named Desire* here, or the time Elaine Stritch was spotted wandering around Town Hall in a bra and slip while in town for a show. "It's a crazy quilt," he muses. "I always say Provincetown chose me, I didn't choose Provincetown." And although he seems humbled by the fact that from Mr. Mailer's table one could touch anyone in the world, he says what he loves is anyone in the world, he says what he loves is that there is no star-driven caste system here. "The attitude is, *So what?*" Now, three decades since this pilgrim first landed, he talks about the intangibles—like waking in the middle of the night to a wine-red sea and an orange moon over the harbor, the perimeter of the scene like a great amphitheater—and yet how substantive it is. "It's like an onion, there [are] so many layers."

We walk across the street to meet up with fellow art scene denizen and D. H. Lawrence scholar Chris Busa. He is, as Mr. Borkowski would put it, a "hip straight" living in P-town. Just as Lawrence went to Cornwall, Chris has returned to this curl of land where his parents once lived, in search of a safe haven. Since the age of seven, he knew Mr. Mailer as a neighbor and friend. Mr. Busa's father was a respected member of the abstract scene (even serving as Jackson Pollack's best man) who lived here before getting divorced and moving to East Hampton, Long Island. Chris grew up watching Mr. Mailer, who sometimes rented a studio space from his family, and clearly paid attention. "You saw how a real pro works," he says, noting how Mr. Mailer was not distracted by things like German television crews showing up to interview him. Rather, he seemed to relish it. The friendship shaped him, he says. As we sit in Chris's living room, art is all around us. There's a Miro hanging over the stairs that's very distracting to me; it beautifully evokes the photisms that I and many other synesthetes often see. Artwork by his talented father, with its bold brush strokes, envelops us on the walls around the sectional. The paintings seem to fairly hum with color.

Chris—the man Norman Mailer once said was too handsome to be a writer—is now not only writing but publishing a journal about the arts scene here, the *Provincetown Arts Press*, and, despite a brutal print market, ad revenues are pretty stable. He points to a two-page Marc Jacobs ad in his magazine as proof of health. He says he's further developing the magazine's website. It has to work. "I don't have a plan B," he says. In the beginning he traded tennis lessons for rare poetry books and sometimes even got paid to earn a living. Now, in addition to the ad revenue, there are grants. "I love doing this," he says. There is a density of knowledge in P-town that seems to work for him. He says that although much of the personal experimentation that was an early draw to P-town is now not verboten in other places, there is still a vitality here, and people can be affected by the energy. He talks about how he is learning to read paintings like poems, and how he is trying to paint a verbal picture in memoir. To me, his language transcends metaphor; somehow I know he's being literal and cross-sensory, and so I ask, "Are you a synesthete?"

He says he is.

His gift seems enhanced by the fact that he can't see out of his right eye. The blind often experience synesthesia; one famous example is Stevie Wonder, who gave one of his albums a synesthetic title, *Songs in the Key of Life*. The cross-sensory joinings central to the phenomenon may have been even further enhanced for Chris in a compensatory way, as hearing and the other senses assumed greater importance. Later, at a party at Dr. Lennon's home, he says he believes my ability to spot other synesthetes is a special gift. Synesthetes, to me, have a special light around them, as he does, and their word choices are often an enormous clue, as well.

I have learned here, through Mr. Mailer's example, what it means to be a New Journalist—to add the observer's perspective into the non-fiction mix, and how powerful that can be. My classmates and professor all urge me to make the writing sample I've brought more personal, to essentially function as a tour guide for readers trying to understand such an ineffable subject. It is at first an uncomfortable position to be in. Traditional journalists like myself are trained to stay out of their stories, not insert themselves into the narrative. But I agree with them that synesthesia requires such a perspective: Because it is such a difficult thing to describe, a narrator must add explanations along the way.

—

Not long after I arrive home, Dr. Lennon contacts me with some electrifying news about Mailer intimate Marilyn Monroe. The two-time Pulitzer Prize–winner, of course, wrote a controversial biography of the incandescent star, *Marilyn: A Biography*, in 1973. Dr. Lennon was reviewing this biography during his research for his own book when he came across the most remarkable find between its covers. According to Mr. Mailer, Marilyn Monroe was a synesthete. Marilyn's first husband, Jim Dougherty, surmised that the reason for the split between Marilyn and her second husband, Joe DiMaggio, was her inability to cook. But there was more to it than that:

> [H]e recounted evenings when all Norma Jean served were peas and carrots. She liked the colors. She has that displacement of the senses which others take drugs to find. So she is like a lover of rock who sees vibrations when he hears sounds, and it is this displacement which will keep her innocent and intolerable to people who hold to schedule. It also provides her natural wit. Ten years later, when reporters will ask her about the nude calendar pictures, she will reply to the question, 'Did you have anything on?' with the answer, 'Oh yes, the radio,' a quip quickly telegraphed around the world, but just as likely she was not trying to be funny. To lie nude before a photographer in a state of silence was a different condition, and much more naked, than to be nude with the protection of sound. She did not have a skin like others.

What Mr. Mailer described was, admittedly, Marilyn's eccentricity, but it was also her synesthesia. Her love of colors, even in food, is the first clue, and the "displacement of the senses" that others use drugs to attain is the most certain one. Her wit and her "skin unlike others" more than confirm it for me. By passing along this news, Dr. Lennon has contributed brilliantly to the history of the gift, and to my personal inventory of it.

I'm so excited by this discovery that I race to the Strand bookstore in Manhattan to get my own copy of her biography. I need to hold it in my hands; I need to own it. A salesperson there climbs a tall ladder to retrieve the oversized coffee-table book with the magnificent photography for me from an upper shelf. I quickly flip to page forty-seven and there it is, on a right-hand page. I can barely breathe as I pay for it at the register. I already know what this will mean to thousands of synesthetes and researchers around the world. Not since the *fin de siècle* celebration has this mysterious trait had such glamour attached to it. Though Mr. Mailer did not call it synesthesia—who did in that era?—it is clearly the gift he is speaking of when he describes her this way. It proves to me that Mr. Mailer had an intimate understanding of the gift, that he would include that detail in her biography.

When I finally collect my thoughts, I know I must reach out to Marilyn's surviving relatives to find out more about this revelation. After much research, I learn she has a niece still living in North Carolina named Mona Rae Miracle. I love her improbable name; and because the whole thing to me is such a miraculous find, it seems apt, too. Along with her mother, Berniece Rae Miracle, Mona wrote a well-received book on her aunt, titled *My Sister Marilyn*, which challenges many of the sensational rumors surrounding the incandescent legend. It takes quite some time for her to respond to my email, but when she does, it's the confirmation I was hoping for, despite the fact that they did not call it by its clinical name in those days:

Synesthesia is a term Marilyn and I were unaware of; in the past, we simply spoke of the characteristic experiences with terms such as "extraordinary sensitivity" and/or "extraordinary imagination." I don't know what Marilyn's IQ was; mine as tested is in the top 3 percent of the population (not genius). Marilyn and I both studied acting with Lee Strasberg, who gave students exercises which could bring us an awareness of such abilities, and the means of using them to bring characters to life. As you know, the varied experiences can bring sadness or enjoyment. I particularly enjoy manifesting odors. Marilyn's awesome performance in Bus Stop (the one she was most proud of) grew out of the use of such techniques, and quite wore her out.

To think that Marilyn's synesthesia informed her amazing performance in that film, and likely those in countless others as well, is astounding. It's also another testament to the large number of synesthetes in the arts. I share the news with my network of synesthetes and researchers, and am honored when the man who wrote the book that started it all for me, Dr. Richard Cytowic, asks to include this information, along with anecdotes from my interviews, in a prestigious lecture at the Library of Congress. "I can't speak of famous synesthetes without mentioning Marilyn Monroe," he says during the lecture, the rich wood paneling behind him adding even more luster to the respectful occasion at the library in the nation's capital. "*Now* I've got your attention," he teases, his audience chuckling knowingly. "A kiss on the hand may be quite continental, but colors were Norma Jean's best friend."

—

In Provincetown, I learn the value of inserting my point of view into a work; but to get truly comfortable with this process, I know I must find some people who have more objectivity about synesthesia to interview, as well. I think that the purest form of this objectivity would have to be found in people who have experienced life both ways, both with synesthesia and without. . . .

6

A CRYSTAL CLEAR PRISM

I'm in Michaels craft store. There's a lavender ribbon here, with chartreuse polka dots that bounce all over the white space from the shelf when I look at it. I feel like I should almost duck going past it, lest the comet-like green balls pepper me like buckshot. It's kinetic—the dots seem to dance on its expanse, and though I can't really think of a use for it at first, I must purchase it rather than leave it behind singing children's songs as if from the back of a car, riotously, road-trippy, there on the shelf, bothering the loudly snoring green velvet, the white eyelet strands with Connecticut lock-jaw, and the plotting, punk magenta tulle rolls. Synesthesia inspires a deep love of color, even imbuing colors with personality. Would I be sad if these impressions were to disappear? I've never known life any other way. And so I think: What would it be like to know life both ways?

Noted novelist and artist Douglas Coupland (*Generation X: Tales for an Accelerated Culture*) is beside himself. He can't seem to make it down a colorful aisle at a local crafts supply shop. He acquired his synesthesia through a health problem, but it puts him emotionally over the top, in a blissful way. He writes me from Vancouver, British Columbia, to tell me how it feels to relate to color this strongly, and I immediately understand. It's paralyzing, he says. He's adjusting to the extrasensory input in an already creative mind. To develop synesthesia later in life is like not having enough melanin in your skin, I imagine, and being dropped into the sunniest of climes.

Across the world, modern philosopher Dr. David Chalmers, author of *The Conscious Mind*, and known for coining the phrase "the hard problem of consciousness," gets on the phone from Australia.

Though I'm trying to find out if he believes synesthesia is central to his area of study, he coincidentally admits that he had synesthesia until he was twenty. He outgrew it, as some people do. He still has the diary entry lamenting the day it disappeared. He took it for granted until it was gone, he tells me. I often took mine for granted, as well, and can't imagine how it would feel not to have this overlay of color and impressions constantly informing my reality.

Euphoria.

Regret.

The pendulum swings for me between these two modern geniuses, each of whom changed modern intellectual discourse in historic ways, even coining new terms strongly associated with both of them. They are important to consider because they are the rare people who know what it is like to live life both with synesthesia and without it—something I can never explain from my vantage point. I'll always be somewhat subjective in that sense. For me, it is like breathing: I was born with the lungs that support these altitudes; I wasn't suddenly thrust into stratospheres of perception that I couldn't handle, nor have I ever had trouble breathing, as Mr. Coupland has. I have never had the experience of waking up one day to find that the fireworks no longer emanate from my stereo when I play my favorite song, as Dr. Chalmers did. These two great minds provide a unique glimpse, and, I believe, one of the only objective ones, into life both ways. The Mailer Colony taught me how important the observer's role can be in nonfiction, but my journalism experience also tells me that the siren call of subjectivity should be avoided at all costs. In talking with these two intellects, I hope to be able to provide what my limited point of view on its own cannot.

If Mr. Coupland's experience is any indication, when synesthesia begins for the first time in a mature mind, it's a joyful but often overwhelming experience. When it leaves abruptly, as it did in Dr. Chalmers's case, there is a sense of sadness, even grief, regarding its departure. To me, this is a big indicator that synesthesia is always a net positive in a human life. So what do these great thinkers make of this curious way of thinking and being? Chalmers has argued that modern philosophy and science do not take consciousness into consideration; that both areas of inquiry fail to account for consciousness at all. This is called *physicalism*. He proposes instead a dualistic view, called *naturalistic dualism*, which recognizes that we are all having conscious experiences. We are aware of ourselves and the nature of the world around us. His 1995 paper, "Facing Up to the Problem of Consciousness," published in the *Journal of Consciousness Studies*, coined the so-called hard problem that is essentially this: Why does the feeling that accompanies sensory awareness exist? He points out that other problems regarding how the mind works, which are not easy at all, are still easier than this important question. And he famously offers his counterpoint to conscious beings, which he calls "philosophical zombies." These are not the zombies of Hollywood, however: They do everything we do, yet remain completely unaware of their experience and therefore stand in contrast to us, we who know what it feels like to hear the sound of a violin or to hold a book in our hands.

It could be said that Mr. Coupland, a postmodern writer best known for *Generation X* (a book that humanized a generation of people coming of age in the late 1980s through meaningful conversations between and storytelling among characters representing that demographic), argues against this "zombification." In writing his book, his hope was to show the group as individuals, not as one mindless mass. He's since lamented that his book has subsequently been misused to categorize and hence further dehumanize Gen Xers, but in the beginning it gave voice to a generation that had previously been lumped into the same group as the Baby Boomers, despite their very different experiences. "I just want to show society what people born after 1960 think about things. . . . We're sick of stupid labels, we're sick of being marginalized in lousy jobs, and we're tired of hearing about ourselves from others," Mr. Coupland told the *Boston Globe* in 1991. And just as his characters at first inspired and energized a national recognition of this emergent generation (not to mention a car, the 1995 Citroen Generation X), Mr. Coupland finds the world fresher and more alive because of his synesthetic perceptions, which he has had for only half of his life.

"The most beautiful place on earth for me is the ribbon aisle at a Michaels craft superstore. I'm serious," he writes. (He doesn't get on the phone; he's phone-averse. He amusedly tells me there's only one in the kitchen and he likes to stare down its handset as he walks past it.) "They have white floors and shelving and cold white lighting, and the proportions of colors on the sides as I walk down the aisle is paralyzing. Try it. But I think it might just be me," he tells me. Well, I have tried it, and he's right. To walk down that aisle, I feel as though I need earplugs or sunglasses, maybe even a flak jacket. Mr. Coupland might be better off trying it with a cane, because the influx of emotions and sensations actually makes it difficult for him to walk. But if it is a paralyzing feeling for Mr. Coupland, it is not one borne of trauma, but rather the weak-in-the-knees feeling one gets from desire.

"For me the [feeling arising from] color is something akin to love. Can there be a link that way? I get crushes on colors—a sort of blissful paralysis. When I visited Ireland the first time in 1995 I found myself immobilized by a kind of green they paint their doors and some of their walls, a shade that never really gets used in the New World. Maybe you know it. It can shut me down. Or a really brightly painted red door. Cornflower blue can stop me. I feel almost giddy even writing these words. So I don't know what you'd call this kind of link. . . . What happens, though, is if these colors join with a certain palette of shapes—then my brain explodes," he writes. "I have this project I've been working on for years now. It's called 'The Brain,' and it's what I tell people the inside of my brain feels and looks like. It's a cube that will be 8' x 8' x 8' and equally dense within, built of certain forms and colours that express time, space, colour, and dimension." He created an art exhibit based on this idea with Medelon Koolhaas when she was in Vancouver. "She has the same sort of pathology," he points out. He sends me pictures that look like songs, full of color and shape.

Mr. Coupland's beautiful love affair with colors rings true for me. Synesthetes will take great pains to describe just the right shade of blue for a photism, and they tend to enjoy color in the physical environment, as well. It's seldom just *blue*. And his enthusiasm for these colors and experiences renews an appreciation for synesthesia in my own heart, making me embrace the condition more, if that were even possible. Synesthetes do sometimes take their gift for granted; sometimes we have to, if only for the sake of getting on with life and living in the non-synesthetic world. But through this string of emails, I'm suddenly buzzing inside and newly aware of the sensations that I've always been privileged to experience.

I first get the idea that Mr. Coupland might be a synesthete when my friend Lisa Fricker sends me a link to an article about him in *The Guardian* by Decca Aitkenhead. The story begins with the startling description of Mr. Coupland sneezing out a grape-sized tumor through his nose twenty years ago. He later revealed in the interview that he responded very differently to his environment after that event. He found himself suddenly mesmerized by colorful wrapping paper and open cans of paint, which he suddenly found to be almost "edible" in all their optical glory. The term "synesthesia" was first suggested to Mr. Coupland by the physician who took care of him after he discovered his tumor. "Am I a synesthete?" he muses aloud. "I suspect so, but my overlaps may be unique. I'll describe it briefly and you decide. Basically my brain sees no distinction between spoken words, their written forms (Japanese as well as English), and shapes as you find in the everyday world: trees, manmade objects, clouds—it all sort of melts together into one single global experience. [Writing] feels no different to me than [making] something. To speak feels like building a house or like using Lego. Using Lego feels like writing. It's been eighteen years of creative work which, over the past year, precipitated a crisis where I had to figure out what was going on in my head. I do all sorts of creative acts for a living and people are always saying, 'Is it weird to go from one mode to another?' and there is honestly no textural difference." It certainly sounds synesthetic to me.

Acquired synesthesia is extremely rare. It is typically caused by injuries to the brain such as a stroke or tumor. The more well-known cases have seeped into fictional portrayals of synesthetes in literature and popular culture. In her presentation "Images of Synesthetes in Fiction," author Patricia Lynne Duffy, an expert on synesthesia in literature and a synesthete herself, has said that a character's synesthesia is sometimes shown as a pathological condition related to brain injury. For example, in *The Whole World Over* by Julia Glass, a character named Saga sees words in color after she experiences a head trauma. "The word would fill her mind for a few minutes with a single color," Ms. Duffy points out, "not an unpleasant sensation but still an intrusion. . . . 'Patriarch: Brown, *she thought*, a temple of a word, a shiny red brown, like the surface of a chestnut.'"

In addition to brain tumors, strokes, and injuries, synesthesia sometimes manifests in people with Asperger's and autism. As of this writing, there are studies currently in the works in Berkeley, California, attempting to find a connection between bipolar disorder and synesthesia, as both conditions are strongly linked to creativity. However, it is important to note that although all of these illnesses are significant challenges, synesthesia does not seem to make any experience, even illness, more negative. In fact, it is often viewed as a positive side effect. Also, synesthesia can exist independent of illness and usually does.

I like the fact that Mr. Coupland is worried about being too precious about it. "I think everyone likes to imagine themselves as special, so I've been hard on myself ensuring that this isn't delusional," he says. "I think that everybody is, in some way, synesthetic; [it's] just that some people have more obvious wiring," he writes. He mentions that he just finished reading a biography of educator and philosopher Marshall McLuhan, the author of *The Gutenberg Galaxy: The Making of Typographic Man* and a fellow Canadian "who was borderline autistic in some areas, and had synesthetic links between graphemes, phonemes, and the portion of the limbic system that regulates the mystical experience. It's me making the diagnosis here—he's long dead—but it's all there." I agree that this could be possible, especially in someone with autism. Synesthesia is believed to occur fairly frequently on the high-functioning end of the autism spectrum. And further, Mr. McLuhan inhabited the arts (he was an English professor), something synesthetes are about eight times more likely to be involved in.

Dr. Chalmers still enjoys music a great deal, despite the missing colors of his youthful synesthesia. In fact, he's been known to take to the stage and sing the blues (and why do we all agree that the blues are, well, *blue?*) at consciousness conferences. "But if I reflect, I do at least miss that extra dimension [of synesthesia]." He found his journal entry the day the synesthesia died and shared it with me in a later email exchange. "Do songs have colors anymore?" wrote the twenty-one-year-old budding philosopher. "They always used to. Most of them were infinitely subtle shades of olive, green, brown (but I could tell). The odd (orthogonal) exception: 'Here, There and Everywhere' was a deep red; 'Ammonia Avenue,' a pure light blue; 'Hotel California,' [a] dark blue. Have all the colors gone now? I took it for granted at the time."

People such as Dr. Chalmers, who lose their synesthesia in young adulthood, are still somewhat mysterious to the medical community, but many such cases are in the literature. "We can only speculate as to why synesthesia might be widespread in childhood but be lost with age as juvenile brains mature," write Drs. Richard Cytowic and David Eagleman in *Wednesday is Indigo Blue: Discovering the Brain of Synesthesia.* "Because it is invariable and overly inclusive as a mode of thinking, it may be that synesthesia is simply replaced by a more flexible mode of cognition, namely, abstract thought and language." When I ask Dr. Chalmers what synesthesia means to him, he considers it from many levels. "First, as a philosopher, I'd first have to ask, What do you mean by, 'What does synesthesia mean?' There are lots of different questions here: Why does it exist? What is its subjective meaning for people who have it? What can we as theorists learn from it?" To answer the first question, he says, "I suppose there is some evolutionary answer that I don't know. For the second, I suspect you'll get a lot of different answers from different synesthetes. For the third, I think we are still figuring that out."

He recalls that he never really thought much of his music-to-color synesthesia at the time. "With the music I'd listen to I'd experience it as having a color. I don't think I talked much about it, it was sort of this curiousity." I ask him if he didn't bring it up because he assumed everyone felt that way. "I think that's exactly it, actually. I think at least for me, I thought this must be totally normal." He explains he didn't know it had a name at the time, like so many synesthetes of his generation, but later he learned about it and thought, *Huh, that must be what it is.*

Synesthetes are usually excited and relieved when they are finally given a name for their often strange and distracting experiences. Dr. Chalmers discovered one of these grateful subjects by talking about the gift in one of his classes, he says. "She came up to me afterwards and said, 'I have that. All these years I thought I was just weird.'" And he met another woman with synesthesia at one of the regular University of Arizona consciousness conferences (where I would eventually meet him later in my journey). "She said she had to compose her own music. None of the other music worked for her. Unfortunately, her music didn't work for anybody else!"

I'm wondering if synesthesia, by nature of its exceptional qualities, could be helpful in getting at the "hard problem" of consciousness as put forth by Dr. Chalmers." Well, I guess every type of experience generates the problem of consciousness," he explains. "The synesthetic experience does, too, but I'm inclined to think that there's already a hard problem thinking about ordinary visual experience, auditory experience. [W]hy is this there? And in the first place why does it feel like something from the inside?" He says he recently had a talk in his Australian philosophy department about this very thing. "The question was, what is synesthetic experience about? [If] you have a synesthetic experience of green produced by a letter or a kind of music, is it really about the color or about the music? Does it represent the color the way a visual experience would, or is it really projecting color into the music[?] [O]r should we say it's really about the music? There's one sense in which it's about the color, but maybe more deeply, it's about the music." Dr. Chalmers has also had the pleasure of being described synesthetically by a student who had person-to-color synesthesia. "I asked her what color I was and she said, 'Oh, you're purple!'" When he asked her if the letters of his name inspired the color she saw, she said that it was more the character of the person and that it could change over time. She added that purple was a good thing.

Both men agree that synesthesia is something to be enjoyed if you "get" it, and something to be missed if it fades away. It's an enhancement, a wonder. Through their fascinating testimonies I have renewed appreciation for and objectivity about the trait. As if to confirm their mutual conclusion, the wonderful *New York Times* reporter and author Jennifer Mascia (author of *Never Tell Our Business To Strangers*) tells me she had synesthesia in childhood and regrets outgrowing it. She remembers the bothersome but fascinating waves of blue and red before her eyes whenever the laundry buzzer would ring. She also used to assign genders and colors to her words. "The laundry bell experience happened when I was two and a half. As the buzzer continued to sound, I remember red and blue blotches appearing in my field of vision, drowning everything else out. In terms of colors, I felt from a very young age that red, orange, purple, and pink were feminine, while yellow, green and blue, white, gray, and black were masculine. As for numbers, 1, 3, 4, 5, and 7 were masculine, while 2, 6, 8, and 9 were feminine. I have not thought about this stuff for so long!"

When Jennifer was a child, she says, she assigned pictures to words, as well. She tried to explain it to her mother, "but she was thoroughly perplexed. For instance, once I said 'damn' and she suggested I say 'darn' instead, and when I thought of her new word I pictured a banana. It was so strange. When I told her that 'darn' was a banana, she thought I was speaking another language," she describes. "I picked up many curse words from my parents. They were from Brooklyn," she adds humorously. Jennifer feels that synesthesia can be frightening for a child, but is more positive for an adult. "[A]s an adult, most would think of it as akin to an acid trip. Many adults are into mind expansion, what with our cultural tendency to dabble in drugs, and if that can happen naturally, it is something to be envied."

Synesthesia researcher and author Patricia Lynne Duffy admits to some negatives with synesthesia, but agrees that the overall effect is positive. "People often ask if there is a 'down side' to synesthesia. Synesthetes often report such experiences as right-left confusion, a poor sense of direction, extreme sensitivity to noise—and I confess to experiencing all of the above," she says. "But even if it is proven one day that these drawbacks are, in fact, linked to synesthesia, I would still never want to lose my synesthesia. I would not trade my synesthetic perceptions even for a good sense of direction, a better sense of right and left, or less sensitivity to noisy environments. I value my colored word, number, and time perceptions. If a bad sense of direction is the price I have to pay for them, then I'll pay it. And I have a feeling many other synesthetes would say the same [thing]."

—

It's this emotional component to synesthesia—the sense that it is positive and something to be missed if you've lost it or envied if you don't have it—that fills me with wonder. Anything that is a considered desirable by people who know life both with and without it must be a gift indeed and not just an idle brain tick. This is a tremendous comfort to me and dispels any lingering misgivings about being "different." As a result, many scientific, clinical views of the trait leave me cold, but if I am to be truly objective, I know that I must nevertheless take a hard look at the history of the science surrounding the trait to see why it is I feel that way. . . .

7

—

SCIENTIFIC SPECTRUMS

I'm in my honors biology class in high school and the formalin vapors (which I can manifest to this day as swirling orange and black and noxious and stinging my nose hairs) have already sent one of my classmates to the school nurse. We are being made to dissect animals. I'm pretending mine is fake to get through it. No, I'm not crying. It's the fumes—really.

Don't get me wrong: I'm very grateful to the scientists who rediscovered synesthesia in the past forty years or so. Without them—most notably Dr. Larry Marks and Dr. Richard

Cytowic—synesthetes would likely have no idea what their unique trait is. The brain scans and batteries and other diagnostics developed by scientists have proven unequivocally that the synesthetic brain is quite a different sort of brain. And recent studies have helped dispel lingering prejudices about the trait even further, by linking synesthesia to creativity and even advanced meditative states.

However, these pioneers and others like them aside, traditional science makes me wary. This is partly because it has allowed synesthesia to fade in and out of awareness according to "acceptable" trends in research. I like to call synesthesia the "Baghdad battery" of brain conditions. The so-called Baghdad battery was discovered (or perhaps rediscovered) in 1940 by a German museum director. It was thought that this mysterious object had been used to electroplate objects 1,000 years before Allesandro Volta's invention of that technology in 1800. Likewise, although both Pythagoras and John Locke alluded to synesthesia in their writings, and although there was previous research (and even a synesthesia-based arts movement) more than a century ago, that awareness and appreciation was lost somewhere along the way. Because forgetting history means that we are doomed to repeat it, modern scientists had to begin anew. That there was so little scientific "institutional memory" in that realm is disheartening. Certainly it robbed synesthetes of my generation of the information they needed to feel secure and even joyful about their gift. How could something so interesting have been forgotten? Was it poor record keeping? A major earthquake or meteorite wiping out a civilization? A devastating fire like the one that burned all the records in Alexandria? Worse: It was forgotten due to shifting tastes and trends in scientific research— something unacceptable for people supposedly deeply invested in objective inquiry into the truth.

Just over one hundred years ago, synesthesia was a blip on the radar of modern science for one shimmering moment before it disappeared. Ironically, the same man most credited for drawing attention to this trait would unwittingly help show it the door. Sir Francis Galton, with his piercing blue eyes and prominent sideburns, defied the conventions of his Quaker upbringing and developed a wanderlust as the wealthy heir of a gun manufacturing family in early 19th-century England. As a young man, he fled from his chemistry studies at university in Germany in favor of touring southeastern Europe. "A passion to travel seized me as if I had been a migratory bird," he said (from a September 1909 article titled "The Progress of Science" in *The Popular Science Monthly*).

Despite his wanderlust and frequent travel, often to Africa, he grew to be remarkably productive. Galton would go on to invent the field of eugenics, create fingerprinting, pioneer behavioral psychology with the study of twins, and further delve into the mysteries of genetics, just as his more famous cousin Charles Darwin did. In his travels, he also discovered this rare ability of people to blend their senses. He would name it *synesthesia*—from the Greek *syn-*, meaning "union," and *-aesthesia*, meaning "sensation." "These strange 'visions,' for such they must be called, are extremely vivid in some cases but are almost incredible to the vast majority of mankind who would set them down as fantastic nonsense," he wrote in "The Visions of Sane Persons" for the *Fortnightly Review* in June 1881. "Nevertheless, they are familiar parts of the mental furniture of the rest, whose imaginations they have unconsciously formed and where they remain unmodified and unmodifiable by teaching." Galton had great sympathy for the synesthetes he encountered throughout his career, particularly when they would relate to him how strange they felt as children. One man in South Africa even sent him a letter enumerating all of his color associations for the scientist to catalog.

Not long after synesthesia made its modest, respectable appearance on the world's scientific stage, a radical shift would occur in the field of psychology that was foreshadowed by Galton's own interest in the psychology of twins. A "second school" of psychiatry would emerge—behaviorism—in which the previous emphasis on inner experience would be sidelined. Galton is considered the father of this school. Later led by American psychologist Dr. John B. Watson, this new school of thought banished inner experience in favor of how people interact with each other. A paper he wrote in 1913 started the wave. Then, in his 1924 book, *Behaviorism*, Dr. Watson explained it further: "Behaviorism [. . .] holds that the subject matter of human psychology *is the behavior of the human being*. Behaviorism claims that consciousness is neither a definite nor a usable concept. The behaviorist, who has been trained always as an experimentalist, holds, further, that belief in the existence of consciousness goes back to the ancient days of superstition and magic" [italics in the original].

The impact of behaviorism was enormous. Synesthesia, perhaps the innermost of innermost experiences, became a forgotten curiosity. The trend effectively obliterated an important finding in a dynamic that is still in existence today. Not until the 1980s, when the cognitive revolution in psychiatry peaked, was it respectable to look once again into internal states. This movement, which had its beginnings in the 1950s, was actually a backlash against behaviorism. It grew from new ideas in psychology, anthropology, and linguistics, and even the new fields of computer science, neuroscience, and AI (Artificial Intelligence). An early figure in the movement was Daniel Broadbent, who wrote the book *Perception and Communication* in 1958, in which he compared thought-to-information processing and used computer terms such as input and output. His model is still in use today. Ulric Neisser, who wrote *Cognitive Psychology* in 1967, said that the mind has a perceptual structure. In the 1980s, the philosopher Daniel Dennett added to the discipline with his thesis that to explain the mind, one needs a theory of content and a theory of consciousness. And finally, AI expert Douglas Hofstadter also shaped the conversation with his wide-ranging interests, including mistakes being a window to the mind and analogy-making being at the root of cognition.

It is against this backdrop of a new, technologically advanced world that two scientists, Dr. Larry Marks and Dr. Richard Cytowic, were inspired to take a deep look inside human experience and rediscover the gift of synesthesia. They would lead the charge that continues to this day in countless learning institutions and labs around the world. But the biases of behaviorism still lingered. It wasn't easy for these early pioneers, who faced skepticism and sometimes outright censure from their peers. Dr. Cytowic said that "colleagues for years refused to accept synesthesia as real and warned that pursuing it would 'ruin' my career because it was 'too weird' and 'New Age.' They had the typical reaction of orthodoxy to something it can't understand— deny it." However, Dr. Cytowic remained fascinated by a case he encountered quite by chance. He knew it was real because this man, his dinner host one evening with friends, kept returning to the sauce he was making to see if it had "enough points." He wasn't referring to the texture of the sauce; rather, he would somehow know it was ready when he felt the familiar ping of "triangle-like shapes" on his tongue. Dr. Cytowic would title his second book, *The Man Who Tasted Shapes*, in his honor.

Slowly, synesthesia began gaining scientific ground once again. This 180-degree turnaround in interest in the trait was the reason Dr. Marks titled his 2003 keynote address at the American Synesthesia Association Conference, "Synesthesia: Then and Now." For years, researchers relied on self-reporting by people who said they were synesthetes, and there was no way to objectively corroborate such fantastic claims. Fortunately, that has changed drastically. The scientific literature now speaks humanely of several generations of synesthetes who spent their lives keeping it a secret, afraid to admit their differences until it became a part of public discourse. Today, universities and labs openly advertise for synesthete subjects on their websites. Prominent artists and scientists have even given interviews about their own synesthesia—a few for the first time in this book. Now that technology and diagnostics have unequivocally proven this to be a real phenomenon, the skepticism and the shame have gone away, too.

Well before neuro-imaging could be used, scientists came up with ingenious diagnostic tools to prove what they suspected to be the case in their synesthete subjects. Dr. Cytowic created several diagnostic criteria to better test people making such claims. The criteria include five points:

1. It is automatic and involuntary.

2. Synesthetic images are spatially extended, meaning they often have a sense of location.

3. The experiences are consistent and generic (the latter meaning simple).

4. Synesthesia is highly memorable.

5. It is laden with affect; it causes an emotional response in the person experiencing it.

Across the pond in London, Dr. Simon Baron-Cohen and his colleagues developed the Test of Genuineness (TOG) in 1987. It measures how consistent subjects' responses are when they hear a sound or look at a number or letter, for example. Synesthetes will almost always see the same color or feel a similar sensation for each stimulus (for example, always seeing a flash of teal when listening to a G note) and therefore score very high, in the 70- to 90-percent range. Non-synesthetes typically score in the 20- to 38-percent range. This test is most effective when it is repeated several months later in order to confirm a consistency in the associations (is that subject's G note always teal, or does it change?).

At the University of California at San Diego, Dr. V. S. Ramachandran and Dr. Ed Hubbard used Stroop interference to create another clever test. They would show the word *yellow* written in red ink to a subject and ask that subject to read it. What they found is that the non-synesthetes typically had slower response times. In synesthetes, however, if the color "matched" the subject's own particular association for that word, his/ her reaction time was much faster. In another similar test, the doctors created a field of 5s in which a small triangle of 2s was hidden and showed it to both synesthetes and non-synesthetes. As might be expected, the synesthetes were much quicker than the non-synesthetes in finding the hidden triangle of 2s because their number-to-color associations made the different number stand out. Finally, an even larger study group was found through the use of the Internet. Dr. David Eagleman created the Synesthesia Battery, an online test that takes advantage of the custom color bars on computers so that synesthetes can find exactly the right shade of indigo for their Hs or that perfect persimmon for their Rs. And exacting brain scans can now be done that show locations of the brain actually "lighting up" when synesthetes are stimulated with various sensory inputs.

Finally, it is *real*. Scientists with their inventive diagnostics have made the intangible tangible. Galton and his synesthetes would be amazed to see this very private experience actually quantified and verified in this way. And now that we have such tools, it appears that synesthesia may be more universal than was initially thought. Some researchers believe that it is widespread in infants, although this theory is currently being hotly contested. For example, in their 1988 book *The World of the Newborn*, Daphne and Charles Maurer theorized that "the newborn does not keep sensations separate from one another." Instead, it "mixes sights, sounds, feelings and smells into a sensual bouillabaisse." This theory was based on their behavioral observations of newborns and still remains unproven. However, it is still an exciting and fascinating thing to contemplate that we all may have been synesthetes at one time! Further, synesthesia has been found to be widely present in adept meditators and has even been induced in hypnosis.

Despite all of these marvelous advances (or maybe even because of them), I don't believe that any one

paradigm—neither the arts, nor theology, nor science—
holds *by itself* the answers to the ultimate nature of synesthesia.
It must be an interdisciplinary approach. And I remain firm
in my belief that those looking for answers only on the sur-
faces of neurons are looking for love in all the wrong places,
at least where synesthesia is concerned. Neither "crossed
neurons" nor a "lack of chemical inhibition" fully explains
what is really going on. Nor do I think that synesthesia
is correctly defined as a "blending of senses" alone. That
definition totally ignores the emotional and spiritual component of the experience. It seems that some common
sense is required, as well as greater input from synesthetes, in order to understand the full import of the trait.

In calling for more interdisciplinary cooperation and
input from synesthetes, I realize that, as Gandhi said, I need
to be the change I want to see in the world. I therefore
stand here as not only a journalist, but an activist trying to
expand the lexicon and hence the cause. If science holds
some answers, I know that I must call on major figures in
the emerging discipline of quantum physics and quantum
consciousness to fill in what I believe are some obvious
blanks. This is partly because the photisms I see resemble
subatomic particles, or at least the graphic representations
of them that I've seen. More to the point, however, quantum
physics often veers pretty close to consciousness and spirituality—fields of inquiry that are deemed less "scientific."
As quantum activist and physicist Dr. Amit Goswami says,
"There is a revolution going on in science. *A genuine paradigm*

shift. While mainstream science remains materialist, a substantial number of scientists are supporting and developing a paradigm based on the primacy of consciousness." And because I believe that synesthesia is a form of consciousness (more than just a result of faulty brain wiring or a glitch in the chemistry of neurotransmitters), this is the next logical step on my journey.

Since this writing, I have coordinated the synesthesia workshops at the Toward a Science of Consciousness Conferences in Tucson and in Stockholm, reporting on what I've learned in hopes that the vast spiritual literature about the trait won't be ignored in the quest for the truth. I believe that it is in these noetic and futuristic realms that the truth—and not pathology—will be found. Basically I believe in its magic, and as fate would have it, a famous mentalist/magician with scientifically inexplicable abilities will soon confirm that for me in ways that I could not begin to imagine.

8

—

RED AND
BLACK MAGIC

Over tea and coffee at The Carlyle in Manhattan, The Amazing Kreskin asks me to slide my reporter's notebook across the table. "You see your numbers in color, don't you, Maureen?"

"Yes," I reply.

"Think of any two numbers one through ten," he tells me. After first only seeing them in black and white, as if he's touched some cosmic reset button in my mind, I hone in on my five and my two. "You have them?" he asks, scribbling something out of sight, his intelligent brown eyes dancing.

"Yes."

"Tell us."

"Five and two, blue and yellow," I say aloud to everyone at the table.

He holds up the pad, and there it is: blue five, yellow two.

Much has been written in recent years about the "secret" of opening one's heart, stating an intention, and having the universe then conspire to bring to you what it is you desire. In this journey I have undertaken, I have intended to fill my life with more light and beauty and to consider the wonder of synesthesia as the central mystery of my time here on Earth. That my quest ultimately becomes a spiritual one seems only right, and that the wonders around this noetic trait never seem to end is a true delight. A most extraordinary thing has happened along the way: Not only am I noticing the synesthetes around me more often, but, perhaps even more miraculously, they are finding me, too.

So it begins with The Amazing Kreskin, whom I meet through a friend. We talk on the phone, and I'm immediately struck by his warmth. He is a student of all things psychological and parapsychological, and shows a genuine interest in my project. "Synesthesia is definitely metaphysical," he concurs on that first magical day we talk. Not long afterward, we join with one of his colleagues—one of my most valuable mentors, Dr. William C. Bushell of

Tibet House—at The Carlyle Hotel in Manhattan, a most elegant meeting spot. Dr. Bushell has studied the greatest lamas, yogis, sufis, and other mystics of the world, rigorously and scientifically, for more than thirty years. As a young man he saw Kreskin twice with his parents in the Catskills. It is Kreskin and Dr. Bushell who will share a most extraordinary bond at that first meeting. "We may have used different railroads, but we got to our destination similarly. It was a joy to be with him," Kreskin later tells me. "You understand this, Maureen. I could enjoy being with him without either of us saying a word. Almost immediately, talk about an energy, an aura—I said to my people I felt so much at home. Now, 'at home' suggests a familiarity, you know what I mean?" It is remarkable to have brought these two extraordinary minds together not only in these pages, but in real life. Dr. Bushell, who has a critical eye (having seen the real masters of the world from Ethiopia to India), is as convinced of Kreskin's extraordinary abilities as I am by the end of our meeting.

The Amazing Kreskin, an American mentalist/magician, got that name from the late Johnny Carson and ultimately changed it legally from George Joseph Kresge. He suggests to me at that first meeting that he often sees a great deal of color, too. I'm intrigued by this assertion, so I ask to follow up with another phone call. I'm getting that feeling in my solar plexus, the one I sometimes get when a fellow synesthete is nearby. The Amazing Kreskin eventually tells me he gets color impressions, too. Not with single numbers or letters, but with words and music. For him, the word *magic* has always been black and red, even before he owned the exquisite black and red silk cape inspired by his childhood comic book hero, the crime-fighting Mandrake the Magician. His dear mother's name is yellow, and a favorite movie and ghost story from the 1940s has a blue soundtrack (not at all a metaphorical association).

"It's interesting about color because if someone brought up to me an art form, such as conjuring and magic, I could see black and red. It is so rich in my life that I could feel the color. I have an incredibly expensive black silk cape lined in red, and it just fills me, because as you know from discussions, you not only see color vividly, but you can feel color," he asserts. He sees auras around people's bodies; most often this occurs with his hypnosis patients. He also has an eidetic memory, as many synesthetes do, as well as perfect pitch; he can play and sing more than a thousand songs, often his own arrangements, from memory on the piano. And needless to say, he is an empath of the highest order, literally feeling the thoughts and emotions of others. I tell him he must be a synesthete. He responds by saying that perhaps that is the reason for his fascination with my book.

He remembers far back into childhood this feeling of fascination with and affinity for color. "Color was a naturally integral part of my life, no one put it there."

And he has noticed colors truer than those in the physical environment—the ones in his own mind. "In movies there's also a great richness. Orson Welles taught me this: There's a greater richness in some respects [than what] you can project [onto the screen]. Don't misunderstand me, I like movies like *Gone with the Wind* or *Dr. Zhivago*, but in many respects there's more color to black and white movies. [. . .] [C]ould you have a more natural richness than what comes out of you? The best scenic director or the best technically minded person is never going to hit the exact color that would fit your response to something." He naturally thinks of color in settings that others would not, he explains. "When I sit down at the piano and play a piece like I did the other night to an audience, 'Stella by Starlight' [from] the movie *The Uninvited*—when I play that theme song from the movie, I can see dark, rich blues seeping through the story, that suddenly become darker to become black. It's not a violently bloody movie but it is a ghost story." He tells me that periods in history also evoke color for him. "When I think of the Salem witchcraft trials, a tragic early part of American history, I can see color all around it; dark grays and browns. The more I'm thinking about it, I think, *Oh, my god*."

"Perhaps that's why I've come into your life," I say.

"People who know me well, I can tell [what color they are] within three or four minutes on the telephone. I get an aura, and the bottom line is I just love colors. What a dimension in life to have color everywhere," he remarks. "I'm in awe of a lot of things that we don't often think about. This life is absolutely delectable." What a supremely synesthetic comment!

He has begun to study the trait in earnest. As someone who can speed read three or four books an evening and, moreover, retain the information, he's already gotten to the heart of the matter by the time we talk.

> It's interesting how when you look at the historical factors of synesthesia, going back to Greek times and in recent times when it was totally ignored. We seem to forget. [E]ven though I've been an entertainer since my early teens, my background is in psychology and I spent time with a very fine psychologist from Denmark, Dr. Harold Hansen, who worked in intelligence in the second World War here and then he was a clinician in New Jersey, one of the finest I have ever met in my life. He really got results. He was truly beyond his clinical knowledge. . . . I had this office for years with him and he felt I had a healing quality around me. It's interesting because as the years went by we came to an interesting conclusion, I certainly had less right to do it. But having read a few thousand psychiatric and psychological texts and catalogued them in my own home, he used to say, "Throw them out the window." I understood years later why. If you look closely, the theories of psychology change every eight to twelve years, with new knowledge and a new perspective[s] and new people. So we have to look at the fact [that], in all walks of life, including

synesthesia, which was ignored and then celebrated and forgotten again, and now people are showing great interest, people tend to become authorities on their own point of view. Even a highly trained clinical behaviorist becomes an authority on his own point of view.

This explains the frustrating gap in interest and inquiry into synesthesia so well, I think. "There's a lot of dogma in these things. This is not criticizing any area of therapy but the problem, and we have to be aware of this as fair human beings, it happens in broadcasting and other things, when an organization becomes successful it becomes dogmatic," he explains. "Then when something 'poses a threat' to their thinking, it is often not scathingly denounced but often minimized or ridiculed."

The Amazing Kreskin says he *hears*, quite vividly, the ideas in his head, and can easily pick out the conversations around him—in a restaurant, for example. "If I listen to an orchestra, I can hear every piece of the orchestra, [as though] I were dissecting all the pieces." He is like the great conductor Arturo Toscanini in this regard, who also shared his photographic memory. But despite his seeming omniscience, life can still surprise him. He recounts extraordinary instances of synchronicity among audience members from different nights in his road show, when unplanned similarities occurred twice in recent performances. He admits to me how stunned he was when two women on different nights picked the same word out of a book, and when two men on different nights both mentioned the brothers they had lost fourteen months previously. I tell him I think the universe is conspiring to show him the wonder, just as he has brought that feeling to audiences worldwide for decades. Kreskin tells me after each of his thousands of performances over the years, he writes a page or two about the evening. He has had access to an amazing traveling laboratory over the years. "It's unscripted; my audience writes it for me," he says, obviously delighted by this fact.

I confess to The Amazing Kreskin that when he first asked me to pick two numbers at The Carlyle, I drew a blank. I saw one through ten as black on a white background and struggled to find the colors. The thought passed through my mind, as it sometimes does, that I'm a faker who cannot prove to anyone that my colors really are there. It was the strangest feeling. Amazingly, he says that he knew this was happening: "I know you did at first. You know what's good about that is that it left the tapestry fresh. You're not a faker. You're a *faqir*," he says, alluding to the wandering ascetic Sufi holy men possessed of miraculous powers. And I believe that this truly amazing man has been perceived merely as a showman when in fact he himself is a yogi master of the highest order. "The mind is a fascinating thing. My life has really been an incredible adventure," he says, the true student of life that he is. "I'll say it as they do on television," he tells me at the close of our conversation. "This is not goodbye. This is 'to be continued. . . .'" His seeing the colors in my mind and ascribing a sort of magic to them is far more meaningful to me than any MRI image or online test. I will soon turn my attention to quite a different sort of magic—the magic of a mind that is at once synesthetic and savant.

9
—

BEHIND
BLUE EYES

I wander into a tiny pet shop off Pont Neuf in Paris; it is filled
with the most richly colored avian species I have ever seen. The
parrot there, he is the primordial green of my Saturdays; the cap
of that finch is the powdery blue of both the note and the letter C;
another bird I don't recognize is the scarlet of E but also the plain-
tive cry of the cante jondo song that accompanies flamenco dance.
Is it the birds or is the light different in France, I wonder. For it is
here that the symbolist poets Rimbaud and Baudelaire appropri-
ated their drug-induced cross-sensory experiences for their work;
here that the true synesthetic composer Olivier Messiaen first drew

*a breath, and here that the likely synesthete Vincent Van Gogh
painted with his thick and richly hued oils. And it is in France
that the most remarkable synesthete in the world, the savant Daniel
Tammet, has now taken up residence.*

Brit Daniel Tammet has taken his beautiful mind to fairer environs. As we chat via Skype, a graphic of the flag of France is visible next to his number on the computer screen. He talks with me about this hopeful new chapter in his life from his hilltop perch in the old papal city on the Rhône. He has moved to Avignon in the South of France, a far cry from the working-class neighborhood in London he once knew. There, his childhood was marred by seizures, which he describes in his autobiography, *Born on a Blue Day*. His world was as limiting as the Asperger's syndrome that also challenged him. But now, he has fallen in love with a talented French photographer and, subsequently, with the art-filled, inspiring surroundings. As miraculous as his mind is, it also exhibits a good deal of common sense. With those empathic, sensitive mirror-touch neurons, he is taking good care of himself. If he must live with this body and soul that feel and see so much, shouldn't his environment be nourishing and as gentle as he is? Tammet agrees:

*I think as a child my autism [Asperger's syndrome is a milder form
of this disorder] was obviously an overriding influence; I was
very stunted in a sense that I wasn't so into the world, and that
included the world of art. I was very much in retreat. But since*

*then the evolution has been a result of a lot of work, a lot of love,
and a lot of effort, and today, art plays a big role in my life—all
kinds of art: painting, literature, and so on. Being in France, being
in the South of France, in Avignon, it's a very historical city. The
Popes moved [here] in the Middle Ages for a period of time and the
palace where they lived is still in the city. And I have an apart-
ment in the hills of Avignon. I have a view of the whole region, the
whole city, the mountains. It's very inspirational. It's wonderful
for an artist, I think, as I aspire to be.*

Mr. Tammet has begun painting and drawing his glorious numbers, the ones that dance on a landscape in his mind
in color and movement in such a way that he was famously
able to memorize the number Pi to 22,500 places. And he
has begun writing fiction to better exercise a synesthetic
mind ripe with metaphors. "It's a very inspiring environment
and one very different from the one I grew up in, being born
in a very poor part of working class London. There is a lot
of art here and I visit museums and I go to galleries [. . .]
[I]f you want me to give you names there are so many, but
I could cite the usual suspects like Vincent Van Gogh, for
example, who worked [in Arles] very close to where we live
now. I often visit the museum here where many of his paintings are kept and stay in touch about his work, about his life.
There are some similarities [between us]."

Mr. Tammet says that he is very moved by the parallels
in their lives. "He's someone who had epileptic seizures, like
I had as a child, in his case unfortunately those seizures

continued throughout his life. That may have contributed to his creativity. And he was also a very spiritual person. He was at one point considering, before he became an artist, entering the monastery and dedicating his life to religion." (Mr. Tammet is likewise a spiritual man, whose thoughts will appear in a later chapter, along with my own, on what I believe is the ultimate truth, the mystical aspect of this trait.) I can't help but hear Don McLean's song "Vincent (Starry, Starry Night)" when Mr. Tammet, characteristically soft-spoken and thoughtful, speaks of Van Gogh. I try to imagine Daniel's world, echoing the singer's lyrics, reflected in Daniel's eyes of china blue.

His move to France a few years ago is "obviously a different story from the one I wrote in *Born on a Blue Day*. [. . .] And so much in my life has changed since [then] in terms of my personal development, the development of my ideas, my creativity, and my personal life, as well. Some of those changes [are] obviously positive, 100 percent positive, and some of them obviously are sad endings; always sad in [the] sense that you give something up in order to gain something new. But I'm very happy where I am today." Mr. Tammet is a noted polyglot who speaks French fluently and once learned Icelandic in a week when someone challenged him to be interviewed in that difficult tongue on a talk show there. Now he uses French every day but continues to write in English.

Though he is grateful to science for healing his seizures, his retreat is as much a break from that clinical world (chronicled in two previous books and countless scientific papers about him) as it is from his humble and painful roots. He has been pricked and prodded, hooked to electrodes and even lie-detector tests by well-meaning scientists around the world interested in his special case. "As a young child, I was obviously healed by what science made possible. But I think there is a danger in areas of quantifying, in areas of analyzing, in taking such a purely objective view of something which is also subjective, the experience of having a brain, having a mind. Our thoughts and our feelings, of course, are not wholly objective, they're inherently subjective. And that's the danger, and I think as long as we're aware of it and can push back against it, I don't think that these two views are necessarily incompatible."

We talk about how his second book, *Embracing the Wide Sky: A Tour Across the Horizons of the Mind*, called for science to take a more inward look at people with different minds. The title was inspired, he says, by an Emily Dickinson poem. "What I was aiming to do in that book is to put forth a vision of the mind from the inside out. . . . [W]hat I mean by that is that neuroscientists have hypotheses, have theories about how the mind works and they approach the subject almost invariably according to scientific method—from the outside looking in—to guess about how it might be to be like a savant, how it might be to be creative, synesthetic, and so on, and then apply those theories to the evidence and see how they work. [O]f course this process isn't without flaws [but] what I want to do is make a contribution to the literature and to the debate. . . ." Though he provides a fresh perspective, he is both sensitive to the evidence and personally revealing. "[O]bviously the difference is that I, having synesthesia, and having savant syndrome, can give that internal perspective that scientists previously lacked, more or less. And it's one that is becoming more available as other savants are being diagnosed; high-functioning savants who are able to express themselves, who are successful in their adult lives. And so what I'm interested in doing, or what I did in that second book, was to explore how savant minds work and what makes me different, but also what makes me similar."

He tries to speak to a common humanity in advocating for synesthetes and savants, both of whom "are generally thought of as [. . .] marginal phenomena, [whereas] in fact I think they're central to how we understand ourselves and how we understand the human mind and how it works." He also credits his synesthesia as being the lynchpin of his abilities. "I think that synesthesia not only gives me those abilities, but curiosity about those abilities and how they work, how they vary depending on a person's environment and culture and so on, and biology and the different kinds of synesthesia." Although I'm happy to hear him speak so positively about the gift we hold in common, his abilities are far beyond my own and those of most other synesthetes I know. His synesthesia has combined with his Asperger's and savant syndrome to create something quite extraordinary. Three "deficits" end up creating a kind of alchemy that puts him in a very rare category of ability that has even inspired an eponymous documentary (*Brain Man*). That said, he's loathe to see himself as anything but merely human. "For some reason I really dislike the term 'human computer' or 'human calculator' that people sometimes attach to my story and to [those of] other savants," he says.

He has been puzzling lately over synesthesia's potential tie to the origins of language. "In terms of language, for example, how words that begin with a certain sound are associated with a certain color or a certain phenomenon, like light, for example. English words that begin with G tend to be associated with light, like glow, glitter, glimmer, glare, glass, and so on, and this phenomenon isn't studied often by linguists, but I think that there are linguists who are beginning to perhaps look at that evidence and reassess it and to realize that maybe for a time they were completely arbitrary, as perhaps many linguists hold, and that they actually result from top processes in the brain [and] associations which are synesthetic."

He adds that no one knows where language came from, and the sheer number of languages around the world further complicates the overarching question of its origin. "We can never know for sure, short of inventing a time machine, but the theories are interesting and actually do point to the fact that there is a connection between how we live in the world, [how we] experience the world as human beings (our bodies, our senses, our emotions, and so on), and how that contributes to language. And the idea that the way we speak, the way we communicate, the way that we connect and associate ourselves with one another, has nothing to do with purely genetic or purely intellectual exercises. . . ." The gentle genius doesn't believe that language evolved arbitrarily. "I think that words are not as arbitrary as that. . . . I think reasons involving emotion and so on [contribute to] why people say one word or another. I think what's interesting is when we do talk about words, which are otherwise a combination of letters, they do provide such strong emotions within us, and because we know what dogs are, we can visualize them; [the] word isn't just letters. [Words] have their own lives."

It turns out that Tammet may be right: synesthesia may have contributed to early proto-language. According to Sharon Begley in a *Wall Street Journal* article titled "Understanding Why George Gershwin May Have Called It 'Rhapsody in Blue,'" "Synesthesia may even explain one of the great mysteries of science—how language originated. Try this: Draw one spiky shape and one rounded, amoeba-like one. Pretend that, in a lost language, one is a 'kiki' and one a 'shoosha' ['bouba' is often used for this second word, too]. Which is which?" Ms. Begley points out that almost everyone says the spiky shape is the kiki, while the amorphous shape is the shoosha or the bouba. The obvious conclusion here is that sounds and forms are linked in some essential way that exists outside of meaning. If this is true, synesthesia may offer us unique insight into how language came to be in the first place.

Numbers engender a similar experience in Mr. Tammet. "Different numbers have an identity, and those identities are obviously the result of the way that they relate to other numbers. And the ability to visualize them [. . .] makes it easier to do sums and remember combinations of numbers or perhaps 'landscapes,' the way I describe it: numbers and shapes and so on." He also says his numbers are kinetic. "They're certainly multisensory, there's feeling there, there's movement there, there's color there, [there's] a lot of emotion invested in all of this, and language, as well." He has created several paintings and drawings of numbers, which he describes as a very satisfying and enjoyable experience. To paint numbers is not a still-life exercise for someone like Mr. Tammet, or for me, for that matter. "The numbers are not simply squiggles on a page [. . .] [they] have their own meanings, which [are] a little difficult to flesh out with color and texture, etc., in the same way that words are combination[s] of letters." How he manages to capture them, constantly moving and vibrating and in different corners of the visual landscape, is, to me, amazing. He hopes to do an exhibition in the future.

How synesthesia fuels creativity is another big area of interest for Mr. Tammet. To him, the creative process is all about making connections, which is something that synesthetes excel at.

Obviously I just have a curiosity just from the scientific point of view, the evidence for and against, and the studies that have been done, which are very interesting. I like the idea that synesthesia [fuels] creativity, that it [makes] connections or help[s] [. . .] individuals to make connections, and that those connections can incite new ways of thinking, how we see the world. And that obviously is what art's all about. It's about giving people new eyes with which to see the world; pushing them, pumping them to come out of their old circuits of perception and thinking, to see the world in new and fresh, different ways. And so I think synesthesia is definitely linked to creativity; I can't see how it couldn't be. And in my own experience, aside from just the experience of numbers and seeing the paintings and drawings that come to me, it helps my writing.

He now writes full-time. His first two books have been international bestsellers. At one point, he says, the only book that sold more copies than *Born on a Blue Day* was then-candidate Barack Obama's *Dreams from My Father*.

The proverbial elephant in the room (or on the phone line, in this case) is, of course, the fact that not so very long ago, neither of us would have even had a name for our gift, removed from respectable examination as it was. I ask him about behaviorism killing the value of inner experience and effectively wiping synesthesia off the map about one hundred years ago: "I think that's a valid point. And certainly we still find that behaviorism in scientific studies today; the way some people will talk about the brain today being 'just this' or 'just that'—lots of cells and connections and so on. And I think it's a very impoverished way of looking at the human condition—how humans move, how they think, how they feel, how they experience love. And I think clearly it's a way of doing science that is losing ground now [. . .] but it's still pernicious."

He says he didn't know much about psychoanalysis early in his life, "only by name, really, and that's pretty much [it]. I haven't ventured into that territory very much, but I do know that there are advantages to reflecting on different ways of seeing the world as opposed to one that is purely based on a genetic view of the brain, for example, [or] a behavioral one as you mention. It does make important the study of thoughts and feelings, dreams, and all kinds of association[s]. I think that's valuable to an individual. As a scientific enterprise, clearly it's a different kind of science that's much harder to evaluate. . . ." Though he believes the value of psychoanalysis is an individual one, he imagines for some people it could be salvation itself. "I think that's part of the hostility that traditional science has shown toward psychoanalysis."

As we continue to talk, I notice that it's raining, so it's interesting that he next mentions a piece of music that, to him, evokes the rain. "I love music. I have a fondness for Chopin, and I very much like his *Raindrop Prelude*. I guess you could say that's a synesthetic piece of music because it imitates a storm; how the rain first falls gently onto the roof of the house and how it grows more powerful and all the emotions that accompany someone listening to the rain, and as it gets more and more fierce and grows and gets aggressive. . . . [It's] a very moving piece of music, and I always enjoy listening to it as well as other pieces by him and other composers." The rain clouds have passed in his own life, clearly. What's next for this man of endless fiction from his new home. "Fiction [also] interests me, literature, in a way it didn't used to. I guess this is part of my development [and] getting older; part of coming to France and my life here today. And I read constantly from all around the world in different languages, lots of it in English nonetheless, as it has the widest array [of emotional expression]. I think that's what's interesting for me, to learn about human emotions, the inner life, [that which] is hidden from us. But it is at the center of what happens in all of us and what we try to come to grips with."

He's recently read several Japanese writers in translation and favors Kawa Bata, the first Japanese writer to win a Nobel Prize in 1968. "He has a traditional Japanese life and his style is very subtle, very spare. And his work has been compared to haiku poetry, which I also find very beautiful. And then when you find these—not so much metaphors, as in the way of Western metaphor (as much as I enjoy that and try to do it myself in my writing), but the moving together of two experiences or two ideas which seem completely unrelated and which in a certain context make perfect sense or seem to give us a deeper sense of meaning. Of course it's what synesthesia at its best is all about. One example is when [Bata] compares an empty box to the arrival of autumn. It's a very stark image but in the context of the story it works beautifully."

We talk about the symbolists, Rimbaud and Baudelaire, their poems, "Voyelles" and "Les Correspondences," respectively, and how it is perhaps not coincidence that they were both born in France, where salon culture once celebrated our mutual trait. "[What they had was] a kind of synesthesia that I think is more artificial [. . .] [because] these were poets that were deliberately, almost you could say *contriving*, different connections to see what new meanings could come out of them, so it's different than synesthesia in yourself or myself or that other synesthetes would have, but it's [part of] a spectrum in a sense." I mention that these two poets and their work embodied a time when synesthesia was so desirable as to be imitated, whereas in my experience it was usually just a source of shame and anxiety. Oh, to live in a time when it was held up as a wonder rather than a peculiarity, or worse, a "condition"!

Mr. Tammet agrees. "I think that's a shame and [. . .] it shows [that] there was a time that emotion was not the debased currency that it's oddly remained in science. [A]nd [now] science [has become] so much a part of modern culture, I think to the detriment of the arts. [T]hat's unfortunate, and I think we really need both [. . .] [W]e need to get the harmony right between them." For his part, he says he will do his best to contribute as much to the arts as he has to science. "It's a new path," he admits. For him, the future is full of options. "I'll leave no door unopened. I'm so grateful and I feel the obligation to make the best of this gift and to make the most of the opportunities that it has given me so far. One of the lines from my books is about having respect for different minds, and if I had to have an epitaph at this point in my life, that would be it."

"Not for a thousand years," I say.

—

All synesthetes need Mr. Tammet. He is an ambassador who has done so much to raise awareness about this trait as well as the beauty of all minds, even those that are "different." That he has moved to France to live among the artists and walk the steps of Van Gogh is, to me, a positive omen of a time to come, in which an artistic renaissance surrounding synesthesia will match that of the scientific one currently raging. And it is this newfound hope for a renaissance that will comprise the next leg of my journey.

10

THE COLOR OF LOVE

I'm on my knees, crawling into the tiny side door of a tea house in a garden in Kyoto. The small openings of these structures were intended so that all would have to prostrate themselves before entering, thus equalizing all visitors from the start. More practically, it also meant that the Samurai would have to leave their swords outside. Soon, the scent of tea is wafting, and we are washing our hands to begin the ceremony. I hear a tiny bell and my ears are suddenly alerted as well. This gentle synesthetic ritual is an art that grew out of the Japanese imperative of addressing all the senses and blending them into one total experience.

I don't want to be too hard on science, because admittedly it has had some real pioneers who have bucked the trends, despite protests from colleagues, and contributed to our understanding of synesthesia. In general, however, science has been, at most, a fair-weather friend of synesthesia, embracing it only when the winds of respectability have blown its way. Indeed, as Cretien van Campen, a noted Dutch author on the subject, penned in *The Hidden Sense: Synesthesia in Art and Science,* "Over the centuries, science was really interested [in synesthesia] only twice: a few decades at the beginning of the 20th century and now at the present for a number of decades." It is the arts—including the stylized Japanese tea ritual—that have always been synesthesia's true love. They may have lived apart from their muse for a time, but, unlike science, they have never forgotten. From ancient codes linking music to the movement of the planets to the illuminated color organs of the turn-of-the-century salons, from French symbolist poetry and haiku to synesthetic paintings of music, the arts have been a steadfast champion of the gift through the ages.

Despite the fact that synesthesia is seven to eight times more common among creative people, the artists creating these works were/are not necessarily synesthetes themselves. The gift has certainly had its emulators in that sense. For example, Scriabin is now believed to have lifted his note-to-color associations from Theosophist Madame Blavatsky. And although it's impossible to know for sure whether Ludwig Von Beethoven was a synesthete, he always referred to D-major as the "orange" key and B-minor, the "black" key. Likewise, Franz Schubert described E-minor as a maiden in a white robe with a bow "the color of a red rose" on her chest. Clearly, synesthesia has colored the arts, literally and figuratively, since time immemorial.

Representations of synesthesia in art don't necessarily consist of literal descriptions of what synesthetes themselves experience. In *The Honey Month*, the author, Amal El-Mohtar, speaks synesthetically of Malaysian rainforest honey smelling like "cold, wet flowers tangled in syrup." Similarly, in "Music Pink and Blue II," Georgia O'Keefe painted pastel, arching forms as representations of the music she was listening to at the time. These representations are reminiscent of the real experience in that they use cross-sensory or even cross-disciplinary ideas and images as their basis. "While in neuroscientific studies synesthesia is defined as the elicitation of perceptual experiences in the absence of the normal sensory stimulation, the concept of synaesthesia in the arts is more often defined as the simultaneous perception of two or more stimuli as one gestalt experience," writes van Campen. So it seems that synesthesia in art has to be considered somewhat separately from the neurological kind that science studies.

This mutually agreeable pairing seems to have found its beginnings in the 6th century BC, when followers of Pythagoras first started to surmise that musical notes and numbers might be related. They noticed that the pattern of eight notes repeated itself over and over like a mathematical sequence, and so they created equations to represent the intervals. Later, in 370 BC, Plato stated that the "soul of the world" also had such musical ratios. Contemporaneously, cosmologists started comparing the radii of the planets sequentially. According to Dr. Sean Day, a synesthete and synesthesia researcher, "[these] ratios would emerge with the following sequence: Moon = 1; Venus = 2; Earth = 3; Mars = 4; Jupiter = 14; Saturday [Saturn] = 25. This sequence approximated the Greek diatonic musical scale's ratios; thus, the planets were tied to music, and a concept of the 'Music of the Spheres' was initiated." Two decades later, Aristotle advanced the theory that the harmony of colors is similar to the harmony of sounds, and Archytas of Tarentus gave Greece a chromatic twelve-tone scale. Because Greek works were later exported to and embraced by Europe after the fall of Constantinople, the idea that the harmonies of colors are akin to the harmonies of sounds would inspire even more derivatives that would ultimately find their expression in artistic endeavors.

In search of how these early ideas gave rise to synesthetic expression in the arts, I visit Parsons The New School in Manhattan one afternoon to sit in on a lecture titled "Synaesthetic Perspectives" by Ernesto Klar. Mr. Klar has a background as a composer and is a lover of the visual arts, as well. "I always wanted to work around this idea of space, sound, image, sort of addressing all of the senses in some way," he tells me privately. "And so, that's why [I have this] fascination with synesthesia." He says he likes teaching the class to remind the students, mostly design majors, that though their field emphasizes the visual, "we actually use all of the senses all of the time. And they determine what we see many times. In terms of an aesthetic context, creating and being sensitive to that multi-sensorial perspective, there's a lot that can be explored. . . . In that sense, I think we will see synesthetic perspectives more and more."

He begins his talk by explaining those harmonious relationships discovered in Greece. Soon the young members of his arts class are considering the work of Guido D'Arrezo. D'Arrezo was an Italian Benedictine monk and Medieval music theorist who taught singers to sight-read synesthetically by placing the notes in different positions on their hands. A diagram called the Guidonian Hand shows his method.

As a result, he is known as the inventor of modern staff notation. He is also credited with creating the modern solfège system (do, re, mi, fa, sol, la, ti, do) and using it to teach music. Both of these cross-modal teaching methods—note-to-hand and note-to-word—are considered synesthetic in an artistic context.

The medieval church itself was, surprisingly enough, a haven of multisensory expression. Mr. Klar points this out to his class, showing the great painting, architecture, acoustics, and other sensorial endeavors and delights of that era. A fellow New School instructor, the late anthropologist Margaret Mead, also pointed to the great sensuality of this era in her writings. As Michele Root-Bernstein writes in her book, *Sparks of Genius,* Mead noticed the similarities between the church and the ancient Far Eastern tea ceremony, especially in how all the senses were engaged: "Anthropologist Margaret Mead made the same point about the synesthetic nature of the arts of cultures recent and past. 'Rituals in Indonesia or Africa appeal,' said Ms. Mead, 'to all of the senses, just as also a Medieval High Mass involved all the senses, through the eye and ear to the smell of incense, the kinestheticism of genuflection and keeling of swaying to the passing procession, to the cool touch of water on the forehead. For art to be reality, the whole sensuous being must be caught up in the experience.'"

As the church grew, so did its artistic traditions. From 1170 to 1250, the Notre Dame School of music in the cathedral in Paris (and its composers, Leonin and Perotin) developed polyphony, the sound of harmonious multiple notes played at once and usually with the organ. And for the first time, musical and numerical proportions were being applied to architectural construction. People's expectations regarding their senses were growing more complex. According to national bestselling author Michael Gelb in his book *How to Think Like Leonardo Da Vinci*, "Synesthesia, the merging of the senses, is a characteristic of great artistic and scientific geniuses." Gelb believes that one can heighten all of one's powers of *sensazione* (sensation) by practicing synesthetic thinking. "A simple way to begin is to practice describing one sense in terms of the others." He then suggests exercises such as drawing or sculpting music, making sounds of color, and engaging in synesthetic problem-solving by giving challenges and their solutions colors, shapes, smells, sounds, and textures. So if you want to be more like Leonardo, according to Gelb, all you have to do is cultivate your synesthetic awareness!

In the late 16th century, one of Leonardo Da Vinci's intellectual heirs, artist Giuseppe Archimboldo, took the master's refinement of the senses even further. Along with a musician from the court of Rudolph II in Prague, he designed a new musical instrument by adding colored strips of painted paper to a *gravicembalo* (an existing keyboard instrument) in order to represent the different notes. And not one hundred years later, the great physicist Isaac Newton was still trying to find the correlation between musical notes and the colors of the spectrum. Working from the assumption that there are frequencies common to each musical tone as well as to each color on the spectrum, he tried to link light waves to sound oscillations. He finally established the correspondence, only to dismiss it later. According to Dr. Sean A. Day, a linguist and president of the American Synesthesia Association, Newton decided that "there could be no direct correspondence between musical notes and colors on the spectrum, and it was just a mathematical similarity and amusement." Dr. Day explains that it was probably not until 1739 that Lorenz Christoph Mizler developed his schema for colored musical notes in response to Newton: A = indigo, B = violet, C = red, D = orange, E = yellow, F = green, and G = blue.

The stage was now set for more experimentation with color and music. Perhaps the most famous of such experiments was French Jesuit monk Louis Bertrand Castel's *clavecin des couleurs*, a colored harpsichord. "The aim was to render sound visible, a concert for the eye," Klar explains. The ingenious device had sixty small curtains, behind which were small colored glass panes corresponding to various notes. Its renown grew when the German composer Telemann traveled to see it; he composed pieces for it and even wrote a book about it fifteen years later. In 1742, Castel dreamt of passing light through the colored panes for a more profound effect. Finally, in 1876, Bainbridge Bishop invented the color organ, a device that could be attached to a pipe organ and project colored lights onto a screen in rhythm to the music. It was not until 1893 that the British painter Alexander Wallace Rimington filed a patent for the *clavier à lumières*, gaining the attention of the German composer Richard Wagner. (Wagner himself famously stormed off a stage during a rehearsal of *Tristan and Isolde* because the color palette of the set did not match the score. This was in keeping with his famous belief in *Gesamtkunstwerk*, or "total work," and that opera should integrate all other forms of art.) And nearby in France, symbolist poets Rimbaud and Baudelaire would add to the celebration and understanding of synesthesia with their own work.

In his "Voyelles," Rimbaud expressed his belief (later proven by science) that vowels create the strongest synesthetic impressions of color for synesthetes. He also provided his own color associations for these letters without actually being a synesthete himself. Baudelaire's "synesthesia," which is also now believed to have been a stylistic device and not the result of the real condition, painted landscapes larger than just the vowels of words. He said, "What would be truly surprising would be to find that sound could not suggest color, that colors could not evoke the idea of a melody, and that sound and color were unsuitable for the translation of ideas, seeing that things have always found their expression through a system of reciprocal analogy" (from the *Painter of Modern Life and Other Essays*, by Jonathan Mayne). His "Les Correspondences" embodies that idea:

"Les Correspondences"
Nature is a temple where living pillars
Let escape sometimes confused words;
Man traverses it through forests of symbols
That observe him with familiar glances.
Like long echoes that intermingle from afar
In a dark and profound unity,
Vast like the night and like the light,
The perfumes, the colors and the sounds respond.
There are perfumes fresh like the skin of infants
Sweet like oboes, green like prairies,
—And others corrupted, rich and triumphant
That have the expanse of infinite things,
Like ambergris, musk, balsam and incense,
Which sing the ecstasies of the mind and senses.

(Cambridge University Press translation,
reproduced with permission.)

With the turn of the century looming, artists began to look to the future. So the illuminated color organ would give way to colored projections on the stages of theaters. The most famous of these exhibitions was Alexander Scriabin's synesthetic symphony, *Prometheus: Poem of Fire*, which debuted in New York City in 1915, complete with moving colored lights. The simple color organ had clearly been surpassed. However, it was still a significant part of a series of Color Music Congresses in Europe, which were based on an organ developed by Weimer Bauhaus School artists Hirsfeld-Mack and Kurt Schwerdtfeger. These performances were met with such interest that many believed that these musical devices would one day be in every home. The London *Daily Mail* sponsored an Ideal Home Exhibition in 1939, which boasted light consoles and organs for the home, as well as a seventy-foot "kaleidakon" tower.

The advent of television and film provided many more opportunities for synesthetic experimentation, with the Germans once again leading the way. According to Dr. Kevin Dann, author of *Bright Colors Falsely Seen: Synesthesia and the Search for Transcendental Experience*, German artists began to emulate synesthetes after the case of an entire family with color-to-sound synesthesia became widely known in the early 20th century: "They saw it as a form of privileged perception—the new. It was about escaping from the old and reaching into the world of the new—a new clairvoyance that [was] on the horizon and synesthetes offered that promise of escaping the bounds of the sensible into the super sensible." In the 1950s that sensibility was about to find its way into popular culture. Avant-garde German filmmaker Oskar Fischinger experimented by painting color on film in such a way that it would synchronize with a musical soundtrack. He also invented the *lumigraph,* a device that would enable one to create images by pressing one's hands or some other object onto a rubber screen. It was always used with background music from another source. In 1964, filmmakers licensed the lumigraph for use in the science fiction film *Time Travelers.* Fischinger would eventually be recruited by Walt Disney to help create one of the most dazzling representations of synesthesia on film, *Fantasia.* Fischinger reportedly left the studio after feeling disappointed that the popular film offered only an abstraction of the real thing.

Also in the early part of the 20th century, a sort of proto–Blue Man Group evolved in Germany. Called *Der Blaue Reiter,* or the "Blue Rider," it featured a group of painters, dancers, composers, and other artists who were grounded in Wagner's belief of the unity of art. They in turn were very much influenced by painter Vasily Kandinsky, who painted his "Impression 3 (Konzert)" in 1912, two days after being inspired by a Schoenberg concert. Though it is hotly debated whether Kandinsky was a neurological synesthete, Sean Rainbird, curator of the Kandinsky show at the Tate Modern in 2006, titled *Kandinsky: Path to Abstraction,* said, "My feeling is that he was quite a natural at it. To have painted the largest work he ever made, Composition VII, in just three days, shows that this language was quite internalized." The article, which appeared on June 10, 2006, in the *Telegraph*, was titled "The Man Who Heard His Paintbox Hiss," alluding to the time Kandinsky reportedly heard sound when mixing his paint colors as a child. Ward further points out that Kandinsky had a lifelong preoccupation with the link between sound and color. "He recalled hearing a strange hissing noise when mixing colors in his paint box as a child, and later became an accomplished cello player, which he said represented one of the deepest blues of all instruments." Kandinsky is said to have first noticed his synesthesia, appropriately enough, at a performance of Wagner's opera *Lohengrin* in Moscow.

In the fall of 2009, the Guggenheim staged a retrospective on Kandinsky's work, titled *Kandinsky in Performance*. Kandinsky's "Yellow Sound" painting was the inspiration for Rafael Lozano-Hemmer's project for the series, called "Levels of Nothingness." It featured Isabella Rossellini reading from select philosophical pieces on skepticism, perception, and color. Her computerized microphone was connected to a full stage of rock-'n'-roll concert lights to create a look backward and forward at a world perceived synesthetically. This performance will not be the last in a world where technology now eases the path to synesthetic artistic expression. "Synesthesia is an important key to understand the human mind, especially creativity, because the connection between art and synesthesia can reveal precise aspects of human consciousness with great clarity," Dr. Hugo Heyrman, a painter and synesthesia researcher, writes from Belgium. "Modern-ism created the conditions for a multidimensional pictorial autonomy. . . . In periods of intense creativity, the search for correspondences and complementarities between the senses are increasing. Artists are the tastemakers of our experience of reality. In the arts, 'the synesthetic experience' became something to be shared. . . . The complementary power of art and science are extending our horizon of perception. With the use of telematic hypermedia, synesthetic art forms will become much more sophisticated in their own right."

———

I feel that perhaps the arts have been more comfortable embracing synesthesia throughout the centuries because of an underlying truth science has long been uncomfortable with—the metaphysical nature of the gift. I will soon talk with a musically gifted man who is quite at ease celebrating this aspect of his synesthesia.

11

THE SOUND OF GRACE

In my New York City public high school lunchroom, sometimes a rhythm would emerge as a would-be rapper would begin to alternately bang his fists and slap his open palms on the surface of one of the long lunch tables: boom-boom, thap, boom-boom-boom, thap-thap. Soon, the sound would fill the cavernous, industrial-tiled space as other kids joined in and offered counter-beats, the extemporaneous lyrics ricocheting around the space like silver synesthetic pinballs in my field of vision. I drum my desk in memory of those spontaneous and joyous times.

If it weren't for Pharrell Williams's music-to-color synesthesia, countless stars would have had to find a new hit maker. "I'd be lost," the Grammy Award–winning producer, performer, and entrepreneur says on the phone from London. "It's my only reference for understanding. I don't think I would have what some people would call talent and what I would call a gift. The ability to see and feel [this way] was a gift given to me that I did not have to have. And if it was taken from me suddenly I'm not sure that I could make music. I wouldn't be able to keep up with it. I wouldn't have a measure to understand [it]."

Williams, who spent time as a child in a Virginia Beach housing project and grew up to be named *Esquire's* best-dressed man, can't remember a time he didn't associate music with the colors he sees in his mind's eye. "Oh my God, it's always been this way. But I thought all kids had mental, visual references for what they were hearing." Music and Williams clearly have an extraordinary relationship, something that his musical heirs and beneficiaries—from Madonna and Britney Spears to Justin Timberlake and Snoop Dogg—would all attest to. On an episode on ABC TV's *Nightline* he told an interviewer that when his family moved to the suburbs and he encountered a wonderful band teacher, he knew music was for him. "It just always stuck out in my mind. *And I could always see it.* I don't know if that makes sense but I could always visualize what I was hearing. It was like [. . .] weird colors."

Williams was just seventeen when a song he helped compose, the dance hit "Rumpshaker," went double platinum. That would be followed by earning millions per song while writing tracks for Mr. Timberlake's critically acclaimed album *Justified*, as well as Gwen Stefani's "Holla-back Girl," Britney Spears's foray into adulthood with "Slave 4 U," Nelly's "Hot in Here," Snoop Dogg's "Drop It Like It's Hot," singing on Madonna's "Give It 2 Me," and, of course, his own hit single "Happy." Though his sound is immediately identifiable, he has said that he tries not to intrude too heavily on another artist with his track mastering—he simply imbues it with his talent. Speaking about the Britney Spears song on *Nightline*, he said, "I Warhol'd it. 'Cause she's still Britney. I just put my colors on it."

The young man, whose name is derived from his father's (Pharaoh), and who says he has both Egyptian and Native American heritages, introduced a new generation to synesthesia by naming his third album with N.E.R.D. (No One Really Dies) *Seeing Sounds*. Unlike many of the figures I interviewed for this book, he has been extremely open about his synesthesia in the media, raising awareness and adding his own brand of glamour to the gift. As he told *Remix* magazine, "Some people make music and they see things. The trait is called synesthesia. It's when one of your senses gets more information than what's intended. When you hear, your ears send auditory images to your brain. But some people conjure [visual] images to the sound, as well. That's synesthesia. Sure, my lyrics are inspired by synesthesia," Mr. Williams added. "You ask any great rapper or writer or musician, and they'll tell you their craziest ideas come from the shower or the plane because in both places there is sensory deprivation." Laurie Kennedy, the music editor at *Remix*, was pleasantly surprised by the public's response to Williams's description of his gift, which was a cover story. "I wasn't so sure of the topic at first," she tells me, "but after we ran the piece, we got mail from people who are also synesthetes. It's such a cool thing."

In addition to his music, Mr. Williams has become ever more expressive with color as he has matured. He has designed riotous clothing and accessories for his "Ice Cream" and "Billionaire Boys Club" lines, and has collaborated on six different-colored Hermès bags outfitted with solid gold hardware. He has even designed furniture, doing a surrealistic spin on classic Eames designs. He has designed sunglasses and jewelry for Louis Vuitton. According to a press release by Vuitton spokeswoman Mona Sharf, "Vuitton delights in opening up to the foremost talents of our times. When it commissioned Pharrell Williams to design its debut sunglasses collection, it dared to reveal the personality and the sense of style of the star of the U.S. music scene." His sense of style has even made headlines as far away as the United Arab Emirates, where he has been lauded as a trendsetter, fashion icon, and all-around renaissance man. His T-shirt designs for H&M's Fashion Against Aids range has also proven him to be a humanitarian, which is not surprising, given his synesthesia. He has even collaborated with Japan's Takashi Murakami on artwork that was previously exhibited at Art Basel in Switzerland.

When this renaissance man is not moving in the realms of haute couture and modern art, he rocks a skater boy aesthetic—right down to installing a half pipe *inside* his home. But for all the riches he has acquired and for all his perceived materialism, he is a deeply spiritual person. He believes that color is not just a marketing tool or a form of expression, but a key to his spirituality. And he believes that the synesthetic imagery he sees is a connection to a higher power. "To me, [synesthesia] is the absolute, direct conduit to God and the collective consciousness, the mind, and the spirit," he tells me. "That is definitely the conduit. I believe that it's in us, I believe it has the ability to go beyond us and the flesh. But I believe for the most part because most people [. . .] are raised [. . .] to be very attached to the flesh and think that nothing can happen without the flesh. And so that's why a lot of our ideas, and a lot of the functionality of the products that we create and the technology, doesn't serve us as well as [they should] because we are coming closer and closer to realizing that the mind and the spirit [are] essential—and scientists are coming around to that and they're realizing that as they're creating all this Artificial Intelligence."

Williams believes one day we'll understand that the real difference between brilliant computers and human beings is soul. "This essence or soul [. . .] interacts and moves through the body, which is the machine, in a dimension which is on this planet here and now, which is time and space. . . . They're figuring this all out now." I "get" him, partly because he's so future-oriented. Why do the photisms, the illuminated bits of color that synesthetes see, look so celestial, like stars being born or subatomic light particles? It's difficult to not go off into space when describing this phenomenon, particularly when you are wrestling to find just the right words. I agree with him, whether science today can prove it or not. Williams is certainly a man with a cosmic consciousness.

Williams decides to take me on a journey involving harmonics. "Another thing I think has a correlation [to synesthesia] is this: There are seven basic colors—red, orange, yellow, green, blue, indigo, and violet. And those also correspond with musical notes. . . . White, believe it or not, which gives you an octave, is the blending of all the colors. . . ." There's such conviction and sense of discovery in his voice. I check in with Dr. Sean Day, president of the American Synesthesia Association and a synesthete himself, for his view of what Williams is driving at. "[T]his paradigm is coming directly from Isaac Newton, who, during his career, established but then, in the same breath, dismissed, a system of correlations between light and the musical scale," Dr. Day explains. "However, Newton in 1704, in his book *Opticks*, didn't designate specific notes as having color. Rather, looking at the Pythagorean relationships of ratios, he designated musical intervals, which thus could have any note as their base, as colored. To wit: red = tonic, orange = minor third, yellow = fourth, green = fifth, and blue = major sixth. Please note, once again, that Newton dismissed this, saying that, physically, there could be no direct correspondence between musical notes and colors on the spectrum, and it was just a mathematical similarity and amusement."

It probably was not until 1739, when Lorenz Christoph Mizler developed his schema for colored musical notes in response to Newton, that we had a seven-notes-to-seven-colors paradigm: B = violet, A = indigo, G = blue, F = green, E = yellow, D = orange, C = red. Many people falsely attribute Mizler's paradigm to Newton. According to Dr. Day, "After Mizler and up until today, many others (such as Scriabin) have come along with variants of this paradigm."

Mr. Williams feels an affinity for color associations in many other mediums besides music. "Colors are light in the electromagnetic spectrum. For every color, there is a sound, a vibration, a part of the human body, a number, a musical note. . . . You have all of your chakras, you have the pituitary, the pineal, the thymus thyroid, the adrenal, the gonad[s], and so on [all of which correspond to colors]." He also has a unique way of accessing his musical gift, which is always accompanied by a vision of colors floating by. He likens his creative process to riding a moped—to fire it up, you first have to peddle it a bit. "What gives me [a meditative state] is the shower. When I get in the shower, because of something called sensory deprivation—when one of your senses is being blocked—it allows your mind to wander and be imaginative. So when I'm in the shower, the water blocks out the sound, so it makes me imagine things. And that happens to me on planes, and that happens to me in the shower, and that happens to me underwater. And that's why when I'm around water it's a little bit more of a creative environment."

This may help explain why he named his successful production company The Neptunes, for the god of the sea. (The Neptunes have won three out of ten genre-specific category nominations so far.) "Sometimes when I'm in the studio if I'm by myself, I'll tell the engineer to just leave me for a little while, I'll turn on a loop of rain, and I'll turn all the lights out in the studio, and I'll close my eyes and sit back on the couch, and I have, like, the craziest visions. That's a technique called sensory deprivation, and that's what you are doing when you meditate. When you meditate, you let go of the senses and allow your inner essence to just flourish and bring to you images of whatever [it wants]." Mr. Williams says that this idea of sensory deprivation is rooted in the wise old wisdom handed down through his family, and is something necessary in this technologically laden, modern world. "It's an old tale among African-Americans in the South that calm waters run deep, and if you would just sit still the answers would come to you. It's because we are so distracted by all of our senses that our minds have to keep up with all that's going on with our motor skills. But when a part of your senses shuts down, the mind just does something else. The mind is an antenna."

Mr. Williams, who heads a company called Star Trak Entertainment and likes to give the Vulcan split-finger salute, believes that synesthetes are people of the future, as are those with attention deficit disorder (ADD). "I happen to have a theory that synesthetes and people with ADD/ADHD will rule the world. You want to know why I think that is the case? Because historically, that is the case." His assessment of synesthetes, though it's a generous one, has some basis in fact. Throughout history, possible synesthetes have included such culture-changers as Leonard Bernstein, Duke Ellington, physicist Richard Feynman, Franz Liszt, Olivier Messiaen, and Vladmir Nabokov, among others.

The music and fashion icon says he knew he was a "different learner" since childhood. "I know for me, when I used to read as a child, my processing—I could read when I had to, but for me, when I would start reading, my mind and my imagination would just go elsewhere." He always had to go back and focus, sometimes reading things two or three times before he would "get" it. He was excellent at reading aloud, but for some reason he couldn't retain the information. "I would be reading, meanwhile reading the faces of the other kids in the room, thinking about this, thinking about that, or 'I can't wait to get home because it's Friday and my Mom makes hot dogs and French fries and I love ketchup.' I was all over the place; my imagination ran wild.

However, [people who are like this] are the best directors in film, [. . .] the best music composers, [. . .] the best jazz musicians. Because we're able to catch all those things, we're also able to articulate [their meaning].'"

Williams says that young people who have similar challenges and gifts think "twenty-seven frames per second," and believes there's nothing wrong with them; they just have the ability to retain more information. "It's only a bad thing when you can't focus and get things done. Most of the people who multitask a lot, they have ADD [that's] just been properly channeled." He also thinks that educational efforts should play to children's strengths, and that children with ADD and those with synesthesia both need to be educated differently. "I'm a huge NLP, neuro-linguistic programming, person; I love NLP. And they tell you that somewhere halfway between a child's hobby (what he does), and his interest (what he'll listen to), somewhere in between there is where that kid is going to be in life, and he's going to be good at it. . . . I believe there are four corners to success, and the middle is the sweet spot, which is the child's destiny. To the left is the hobby, to the right is the interest, to the front is the teacher, and to the back are the parents. In the middle of those four points is the child's destiny. . . . I feel that we need to change the curricula around the world. [T]he problem now is they assume most kids are visual learners, but that's not the truth. The truth is, some kids pick things up [through sound] quicker, some kids pick things up visually, some kids pick things up kinesthetically, and there's also [the] olfactory and gustatory [senses]. And there are synesthetes who know more about something by the way it smells or tastes—like, there are people who know what purple tastes like, and it's not grape."

Would he ever consider collaborating with other synesthetic artists? "I would love to. I just think there are more synesthetes in music than one would ever assume." I send Mr. Williams some resources regarding synesthesia, including a link to a terrific lecture by top neuroscientist Dr. V. S. Ramachandran. I'm delighted to read a few months later in *Interview* magazine that they up met in a California recording studio. "People think of art and science as being fundamentally opposed to each other, because art is about celebrating individual human creativity, and science is about discovering general principles, not about individual people," Dr. Ramachandran tells the interviewer. "But in fact, the two have a lot in common, and the creative spirit is similar in both. It's about seeing hidden links, which nobody else has discovered before. And in synesthesia, what's going on is the brain sees all these amazing links, which you and I can barely glimpse."

Could this be why synesthesia feels like a spiritual state—that we are tapping more easily into a connection between the temporal and the infinite? Is the synesthetic mind a broadband connection, whereas non-synesthetes are still using dial-up? Williams remarks, "Well, [Ramach-andran] just kind of watched me make a song and how, as musicians, we react to something being right, how we react to something being wrong, and we go searching for it out of what's seemingly thin air. But for us, we're noticing rhythms, we're catching something. My explanation is that curiosity illuminates the correct path to anything in life. If you're not curious, that's when your brain is starting to die. And discovering, I think, that's what separates us from the rest of the other species. It's that [although] we discover and pioneer, we don't forget from whence we came." (From *Interview* magazine, used with permission.)

I'm newly motivated to find out more about the spiritual links to synesthesia after speaking with Mr. Williams. I'm left with one word on my mind after our conversation: grace. I recall a hymn from childhood while ruminating on this, and realize for the first time how synesthetic it is: "Amazing grace, how sweet the sound, that saved a wretch like me. . . ." Saved by sound? Grace as sound? Although it is one of the most recognizable songs in the English language and I've heard it countless times, I never heard—or saw—it this way before. What did the former slave trader turned clergyman John Newton, who penned the lyrics as a sermon in 1773 and published them in 1779, experience? How did he experience the divine? He was said to have undergone a miraculous transformation into piety after nearly drowning in a storm at sea. Perhaps this is that "oceanic feeling," a term first coined by Romain Rolland, that source of all religious energy. Perhaps this is what Pharrell and I are able to tap into every day. Perhaps this is what overwhelmed John Newton as he was brought from the darkness of the storm into the light of that amazing grace. Neptune, the god of the sea . . .

12

A TECHNICOLOR RENAISSANCE

A hipster friend of mine sees the vintage black velvet Halloween mask of Vegas showgirl proportions I found in an antique shop. She decides to build a theme around it for Rubulad, the Brooklyn warehouse party of legend on New Year's Eve. The night of the party I make my way to the unmarked address and navigate the stairs, ostrich feathers brushing against the wall as I grip the banister. Inside, it is a synesthete's delight: music and food, and art hanging on every wall, and the diversity of humanity—most in elaborate masks—spread over several sagging floors and even

the roof in a most beautiful canvas of multisensory stimulation,
all aided and abetted by some chartreuse green absinthe. . . .

Knowing that the human fascination with movements often comes in one-hundred-year cycles, such as the Arts and Crafts revival currently underway, I decide that it is high time for a second synesthesia renaissance. Though I'll admit to some wishful thinking, there does seem to be a nascent synesthetic movement afoot, with examples in every medium, including computer-generated art that could never have been imagined during the heyday of the last *fin de siècle* celebration. When I reach out to postmodern novelist and astute social commentator Douglas Coupland about his own synesthetic experiences, he remarks that it is "odd that it seems to be so much [of it] in the air these days." I agree.

Some of the loveliest examples are in children's film and literature. Take, for example, the adorable Parisian rat in the Disney/Pixar film *Ratatouille* who dreams of being a famous chef. Remy the rat is not satisfied with his ordinary slim pickings and confronts his father about their scavenger ways. He longs to join the world of humans and their culinary riches but his father warns him that humans are dangerous. Soon, however, Remy is on the counter of a woman's kitchen, peering over her shoulder as she watches the great French chef Auguste Gusteau's cooking show. The chef expounds on how food is like music but that you have to learn to savor it. So Remy experiments with a hunk of pilfered cheese, and suddenly the photisms well known to synesthetes begin to radiate around his left side. He eats a strawberry, and his right side lights up with more illuminated colored forms. But when he combines the foods in one big bite, the synesthesia explodes around him and the music begins.

If the synesthesia depicted in this film rings true, it's because the animation department had some help. "The animation director of that particular segment contacted me regarding it. Now, that was after they had completed doing it," says Dr. Sean Day, president of the American Synesthesia Association (ASA). "But the team wanted my opinion on whether I thought they had accurately depicted 'taste-to-color' synesthesia; they were even more interested in my opinion when I informed them that I myself had this type of synesthesia. They told me that, even though they had already completed the film (if I remember correctly, it was due in the theaters in just a week), they were wondering whether they might want to adjust that scene for the DVD version. I told them that their depiction was just fine, accurate for the purposes of the movie, and that any changes would be, all-in-all, trivial variations."

Dr. Day was also consulted when the producers of the NBC hit show *House* gave a character synesthesia. This desire on the part of producers to represent the trait accurately is a wonderful sign of the synesthetic times. It seems that young people are generally much more aware of the condition than adults. My mother mentions my synesthesia to her neighbor, who then brings up what she thinks is a new topic during dinner conversation with her daughter's family. Her 'tween granddaughters tell her they know what synesthesia is and run to their bedroom to retrieve a copy of Wendy Mass's *A Mango-Shaped Space*, a heartfelt coming-of-age story of a thirteen-year-old synesthete heroine named Mia Winchell. The book has done a great deal to raise awareness of the gift, and author is a frequent guest lecturer at schools, continuing to educate the public about it.

Similarly, my college roommate tells me her daughter knew about synesthesia when she raised the topic at dinner one night. She had just finished *The Name of This Book Is Secret* by Pseudonymous Bosch, the first offering in a YA series about twin synesthete brothers, Pietro and Luciano Bergamo. In this story they join the circus at the age of ten. Through their synesthesia, they develop an encoded communication system that makes them look as though they are psychic. The act grows so phenomenally popular that Luciano is kidnapped. Pietro searches for his twin for decades and eventually finds a clue. There is a newspaper story about another kidnapped synesthete—a little girl named Lily Wei. In my favorite scene in the book, the now-old magician finds Lily playing the violin. When a beautiful stream of colors pours out of her instrument and swirls around him, he says, "You play like someone who sees music. You paint with sound." Pietro tells the reader, "When she finished playing her final note, the sounds and colors kept spinning in my head, as if she had case some kind of synesthetic spell." I wish there had been titles like this when I was younger.

Synesthesia is not just for kids, though. In Dan Brown's *The Lost Symbol,* the bestselling storyteller explores as many esoteric themes as in his other works. But this time, I was surprised to see a scene describing synesthesia. In the secret basement laboratory of the villain, Mal'akh, the light changes color according to the Table of Planetary Hours. From chapter 81 of the book: "Centered over the table hung a carefully calibrated light source that cycled through a spectrum of preordained colors, completing its cycle every six hours. The hour of Yanor is blue. The hour of Nasnia is red. The hour of Salam is white. . . . Now was the hour of Caerra, meaning the light in the room had modulated to a soft purplish hue." And from chapter 86: "As the light over Mal'akh's head began changing color again, he realized the hour was late. He completed his preparations and headed back up the ramp. It was time to attend to matters of the mortal world." Is the villain Mal'akh a synesthete? Perhaps. Here, Brown is actually referencing the work of Girolamo Cardanus (also known as Cardano), who in 1550 developed a system of corresponding colors with flavors and the planets. And as we have already found out, the ancients before him ascribed musical tones to the planets, which came to be known as the Music of the Spheres. There are many other similar, if less-well-known systems like it.

An example of an adult novel with a compelling synesthetic heroine is *Bitter in the Mouth* by the talented Vietnamese-American writer Monique Truong. Her character Linda Hammerick tastes words. When Linda's grandmother speaks to her in one scene she tastes it this way: "That doesn't stop *cannedcorn*me from being your grand*potatosalad*mother, Linda*mint*."

Synesthesia has even crept into anime. The eponymous character from the TV series *Canaan* is a female Middle Eastern assassin who is able to see the emotions and intents of others in color. She also has a superpowerish ability to change sound to smell and color to sound. And the canceled NBC TV show *Heroes* featured a beautiful storyline about a deaf character named Emma who could see sound. Stunning special effects in one episode showed filaments of colored light emanating from her cello or as she played the piano.

In addition to synesthesia of the literary and dramatic sort, I am hankering for some music, preferably some jazz by the piano legend and synesthete Marian McPartland. The then ninety-one-year-old (now deceased) National Public Radio host consents to an interview and then invites me to see her play at Lincoln Center's jazz space, Dizzy's Club Coca Cola. To McPartland, the key of A is bright pink, and the note D is daffodil yellow. McPartland was born just as the turn-of-the-century fascination with synesthesia was fading, so the woman who still delights audiences with her own performances and who has garnered a large and loyal following with her radio show didn't know that her special way of thinking and perceiving had a name until we spoke. "I've never heard that word," she says, asking me to repeat it. "Well, I've always thought of certain keys in certain colors. There are other musicians who do that, too. The D is a daffodil yellow. And the key of A is kind of a pink color, but that's just an idea. It isn't something I think about constantly. It's just sort of a thing I and a few other musicians, a little idea that I have. It's not anything that I talk about or that I think about when I'm playing. I think you may be making more out of it than it is."

Her rather blasé reaction is fascinating to me. I understand why she might think it's no big deal because we synesthetes function very well despite the influx of impressions we receive; they are like background noise. But I tell her that this isn't everyone's experience, which is why it interests me. Further, I mention that this "little idea" she and the other musicians have is eight times more common in those in the arts. "Well yes, I would think so," she says, "because if they weren't in the arts, how would they get the idea of keys, unless they were a sax player. . . . Maybe because I am a piano player, I think it's especially evident in piano players. [Y]ou just made me think more about it."

Her comments make me wonder about people who may be synesthetes but who never had musical training. Would their impressions remain dormant without an outlet and the accompanying vocabulary to go with it? She asks me about my own synesthesia and how I became aware of it. What color is the A of my alphabet, she wants to know. I tell her it's yellow. "That's funny. I wouldn't think that. I would make that pink, too. I never thought of letters of the alphabet being a color, just the keys of the piano. It's very interesting that you regard it with so much fascination. I don't think I've given it that much thought. I never heard that word [synesthesia]. Where did you hear that word?" I tell her my awareness was raised by *The Man Who Tasted Shapes*, written by a top neuro-scientist, Dr. Richard Cytowic, who once told me he used to sit in her audience when she would play Café Carlyle here in New York City. Ms. McPartland says she has spoken about this idea with Jim Hall, a fine guitarist who also has colored music. "He and I were comparing notes, and it seems to me everyone is different. I see the key of D as yellow and other people have a different idea, maybe blue. I've got B-flat as blue. . . ." Synesthetes' individual colors are as unique as their fingerprints.

I tell her that her national radio program has served to educate many people on the history of jazz, but she is humble about her role in this. "Well I don't think those things. If people are getting something out of piano jazz, that's fine. I'm not really looking to educate anyone. We want to have all the best guests we can possibly find. And we've been on the air more than thirty years and hopefully we're going on. We get tons of mail and phone calls. We had a big celebration recently for the fact that we have been on the air for thirty years. That's about the way things are."

Growing up in England, Ms. McPartland learned to play the piano as a child from her mother, but she never had a conversation with her about her long-standing impressions. "Well, my mother has passed away and she used to play piano. This was years ago; I never spoke to her about it. I doubt if she would have known what I was talking about. She played a sort of prosaic piano and she played [. . .] Chopin. Which is how I learned to play, listening to her. This is all very interesting to me that anybody would find this so interesting," the grand dame of the keyboard opines. The following week she takes the stage in her wheelchair and literally springs to life when her bandmates fire up some standards. The lights of the city through the huge wall of windows beyond her grand piano seem to rise from the keys, as they must also for this amazing woman. It's a magic night and I couldn't be happier if I were looking upon the *clavecin des coleurs* one hundred years ago.

To cleanse my musical palate, I decide to see a concert by a fascinating synesthetic young woman, Norwegian punk star Ida Maria. A writer and performer of songs such as "Oh, My God," seen on television's *Gossip Girl*, and ballads such as "Keep Me Warm" and others licensed for *Grey's Anatomy*, this impassioned young woman has had a critically acclaimed year. Her team invites me to hear her play at the

Maritime Hotel's Hiro Ballroom. Watching and listening to her, I cannot help but think of Janice Joplin, to whom she's often compared; her voice is like honey over sandpaper. She's pure energy as she flails around the stage. Her up-tempo songs are full throttle, but her ballads are tender and reflective. I think back on the gentle stylings of Marian McPartland and realize that beyond the usual color associations, there is no one way to typify the synesthetes whom I've been so privileged to meet. Despite their common trait, the range of expression seems infinite.

If we are discussing the arts, we can't leave out *gastronomie*. Another feast for my synesthetic senses is Dylan's Candy Bar in Manhattan. The chain of stores is owned by Dylan Lauren, the entrepreneurial daughter of fashion icon Ralph Lauren. A friend points out the first paragraph of a *New York Times* story one day that seems to suggest that she shares this peculiar relationship with color. "Watching my dad, Ralph Lauren, design clothes when I was young, I was always inspired. I loved the colors; I wanted to eat the color swatches. I feel the same way when I see a row of Polo Ralph Lauren cashmere sweaters or colored shirts," she said. "I've always seen items like these as food or candy. My mom is a painter and photographer and my grandfather was an artist, so I've always been surrounded by creative people."

Upon reading this, I make an appointment to meet her at the Third Avenue flagship store. I arrive a little early to take in the beautiful rainbow hues of the interior. Candy of every shape, hue, and taste lines the walls. They even have watermelon gummy bears, which I decide to purchase. Upstairs at the ice cream bar area she comes in and greets me, and immediately tells me that I must try the pickle bubble gum ice cream. She has one of her employees scoop some into a small cup. I can't imagine anything more synesthetic than combining those two flavors—and it actually works! We go outside to talk under a tree in an adjacent small park. The conversation turns to her love of the color turquoise. Though she doesn't see letters, numbers, or music in color, she confirms for me that she always associated the rich hues of her father's designs with flavors. I feel certain that this lovely young woman has a touch of the trait (it is possible to have synesthesia to varying degrees), and that her chain of stores is clearly a living testimony to her affinity for color. She has even written a book about her love of candy and candy-themed design, *Dylan's Candy Bar: Unwrap Your Sweet Life.*

My Norman Mailer colony mate, talented writer and instructor of writing Sandra Hunter, once told me she believes there's something synesthetic going on with Dame Evelyn Glennie, the beautiful Scottish percussionist who happens to be deaf but feels the music so fully that her performances are virtuosic feats. After watching a few of her concerts, during which she literally feels the sounds of the accompanying orchestra through her stocking feet in order to set the rhythm, I'm pretty sure Dame Evelyn is using some sort of cross-modal pathway. It may not be garden-variety synesthesia, but it is so remarkable that I must learn more.

To me, it speaks to an extremely heightened awareness of the senses that I'm hoping to cultivate more fully in myself.

When I phone Scotland, she is kind enough to share her thoughts. "I know you feel sound kinesthetically," I begin, "but do you think that some kind of synesthesia is also present?"

"All I do know is definitely the senses do come together, and although [I'm] concentrating on a form of touch, [. . .] I'm feeling whether that vibration that is felt through the body is a wide vibration or a fat, broad vibration, or whether it is something that is thin, or something that is tingly or tickly—that type of thing—or whether it is distant or at the forefront. . . . [B]ut I don't really bring in the sense of color to that at all. It doesn't play any part in this musical emotion or interpretation, it really is a feeling with a sense of touch. Or it could be a particular image. These images can change over time. If I'm playing a piece of music I wouldn't necessarily have the same image every time I play that [. . .] piece or phrase. Or the interpretation may change, the acoustics of the hall may change, the instruments may change, and so on and so forth," she says.

What images does she see? I wonder. "It could be as simple as thinking of a past experience, it can be a particular facial image, it could be thinking about a particular emotion, whether it's anger or something passive, whether it's something that's more loving or cheeky. . . . And the images are not so much abstract in my case. It could be something that I literally experienced. It could be a particular memory of a brook or a stream or a river. Or it could be a particular image of a castle that I experienced as a young person—that type of thing, rather than something abstract. So actually everything is pretty simple and quite normal, so I don't feel that I'm overly complicating things. But definitely the visual aspect is so important. I enjoy writing music for the media, writing music for television and for film, because you are connecting to something that's very visual indeed.

"Having said that, if I'm giving improvised concerts, which I do with the likes of Fred Fritz on guitar, then clearly you're starting with a clean slate, a clean blackboard, as it were, a clean sheet. And you're literally painting the sound—you know, using the hall or the room that you're in as your canvas. And in that kind of instance, again, things can be much more abstract. But nevertheless, the idea is simple and the sounds are quite pure, rather than cluttered. It means you're thinking of a kind of emotion or you're thinking of a type of field or a texture, like sitting around on a bean bag [chair], for example. There is the feel of the beans themselves and [. . .] how that feels throughout the body, and there's a feeling that the beanbag is always moving and it curves around the shape of your body. And that I can translate into sound, but I'm not thinking of that beanbag being blue and [thus creating] a particular sound for me. That doesn't do anything for me at all.

"Maybe we all experience this," she supposes. "I don't regard myself as an expert in this at all, I don't really connect [synesthesia] to myself. I guess I don't really want to overcomplicate the things that I do, because it is quite natural for me to do, so I suppose I shy away from over-analyzing things, the same way that when I'm performing, I want to analyze things that will help me improve for the next time, but everything needs to be very simply analyzed so that the goals of improvement are manageable, really. When you think about it, maybe we all have a form of synesthesia. Because we are all really using our senses quite differently nowadays compared to ten years ago or twenty years ago, and so on."

I saw a documentary that showed how certain sounds affect her feet, legs, or stomach, whereas other sounds are felt only on her forehead. I ask her if sound literally moves around her body. "Yes, absolutely," she agrees "Basically I see the body as just a huge ear. It's something that is completely responsive. And the thing about percussion is that we are dealing with many different types of frequencies, different durations, [. . .] different textures, different ways for the sound to travel, [. . .] different dynamics, and so on and so forth. And I think that more than any other family of instruments, this really helps the player to hold the whole body in what they are doing. But also the body, when [a player is] executing a sound through percussion, the body is really pretty open, so the types of postures [are usually] very natural. . . . If you think of a violinist or a flute player or a trombonist or so on, and you think of the position the body has to be in to play the instrument, and you take the instrument away but leave the body in the position it's in, then it's a very unnatural position. But usually for percussion the body is pretty well-centered, and it has a chance to really have that sense felt throughout the body. So yes, absolutely—basically, the low sounds are felt in the lower parts of the body, which allows [. . .] the player to analyze the sound as a fat sound or a thin sound, or [an] aggressive sound, or [a] distant sound, or that type of thing, much more easily than if the body were in an unnatural position. The high sounds are felt in the upper part of the body like the cheekbones, the forehead, the skull, the neck, the breast, and different parts of the hand.

"But of course there is a very different feeling being the participator in the sound [as opposed to] a passive listener. The person in the front row will have a very different experience than a person sitting in the tenth row. And then someone like myself who is actually creating that sound behind the instrument might have a very different feeling than someone who is on another side of the instrument or at the end of the instrument. If you think of the size of a marimba, which is approximately 9' by 5', then heavens, there are so many positions that a player can [occupy] in order to experience that instrument in different ways."

The body as a huge ear is a synesthetic concept *par excellence*. Afterward, I listen to Dame Evelyn's music and I feel parts of it more in my stomach or my legs or my cheekbones. She has given me a remarkable gift in making me more aware of my skin, and I'm deeply grateful. Shortly after our conversation, the *New York Times* reports on a study conducted at the University of British Columbia by Bryan Gick and Donald Derrick that showed that people hear with their skin as well as with their ears! The study also showed how the brain plays a role in sensory integration, and how we must use all of our five senses—sometimes several at once—to interpret the world. I send the article to Dame Evelyn, who is delighted, although of course she always knew this was so.

And what of the painters? The Guggenheim museum is staging a Kandinsky retrospective, so I get myself over to the Museum Mile and enter Frank Lloyd Wright's beautiful nautilus-shaped building, which houses more Kandinsky works than any other museum in the world. As I stand in line among visitors from many nations, I recall how Wright once claimed he could hear music while designing; might he have been a synesthete himself? I make my way up the inner spiral and find myself growing somewhat dizzied by myriad beautiful canvases. But the one that stops me in my tracks is "Several Circles," which is actually a perfect representation of my synesthetic impressions, with its vibrantly colored orbs on a black background. I feel the colors of Kandinsky on my skin and through my whole nervous system. I reach for a post and lean against it for a time. It *is* possible to feel a painting this deeply, particularly when the artist captures an experience that is so difficult to put into words.

—

Fortified by these impressions, I will never again seek the darkness when there is so much light refracting all around me. I'm glad to finally be living in a time where synesthesia is being examined and even celebrated as it was one hundred years ago. Dare I say *renaissance?*

13

—

TASTING GOD

I walk into one of the most important places in Byzantium, Hagia Sophia, Church of the Holy Wisdom, built by Constantine and standing the test of time for 1,000 years, even while Constantinople was being conquered by the Sultans and after the great Eurasian city was changed to Istanbul. I crane my neck to stare up at her giant dome, under which the Statue of Liberty could stand. I see the shimmering mosaics of Christ and the Arabic script proclaiming there is no God but Allah. The sun shines through the stained-glass windows, with colors evocative of my own synesthetic photisms. Yes, this is what they are like sometimes—sunlight refracted through divinely colored glass.

I cannot harness the wind, nor can I put it in a box to turn over and examine. But I have seen kites flying over the Narrows in Brooklyn, and watched the gulls of Istanbul ride its vapors over the Bosporus. This is how it is for me with synesthesia's innate grace. I can't show it to you—it will be visible only to my eyes—but I can tell you what I see and how I feel when I see it, and you can see the effect it has on me. All of those drifting colored forms and sensations are not mere passing impressions; they are "laden with affect," as Dr. Richard Cytowic describes in his five points of diagnosis. My response isn't a passive reflex; rather, it is accompanied by that emotion of certitude, that "aha moment," and it is both outside of me and within me, giving me a feeling of connectedness to Something Else. To me, that something else is God.

I'm long past my days of Sunday school and regular church-going. I don't know if God is a man or a woman or if the Creator prefers certain political parties, lifestyles, or faiths over others. Is God vengeful or forgiving? Or is He/She apathetic? I've sometimes entertained the thought that the Creator doesn't exist at all, particularly on high crime days when no one seems to be minding the store here on Earth. From my front-row seat to carnage and depravity, I believe this is an honest and legitimate doubt. The wisest thing I ever heard about God came from a swami who was touring the States. "What is God, swamiji?" one of the seekers in his audience asked. "My God, I don't know," he replied. I've spent many years going to services that were not my own but the traditions of friends, or even a part of a story I was pursuing. These traditions seem to be but a small part of a wider Godscape than can be imagined in total. In a way, all of them felt like paths leading to one home. I believe in a creator because we all create. In our own overlapping realms, we spend our lives making, doing, achieving, and I think this is rippling out from some sort of unifying creative source or intelligence. My belief is deepened and made whole by the fact that the images and sensations I feel as a synesthete seem to come from some Other Place that I must be tied to somehow.

Yet if I followed a synesthetic filament to its origin, I'm not sure what I'd find. The essential unknowability of the Creator seems to mirror the ineffable quality of my own synesthesia. I can try to pick both of these things apart, examine them, dissect them, but in the end there is a mystery here that, by necessity, remains pure and untouched. But I have to ask the question: During these "aha moments," am I sensing something that's not really there? In other words, is it just me? I wonder if my fellow synesthetes feel this way, too? Therefore in the final leg of my quest, I know I must ask other synesthetes as well as spiritual leaders and scholars what they think synesthesia means, spiritually speaking. My belief that these two areas—spirituality and synesthetic perception—are inextricably linked is such a strong one that it leads me forward.

Although he's not a spiritual leader, there's something about synesthete Manu Katché's drum playing that makes me start with him. Although he's spoken in interviews about his music-to-color synesthesia, he's never mentioned anything about a spiritual connection. However, his distinctive styling on songs such as "In Your Eyes" and "Fragile" makes me think that he has brushed up against the divine right there at his drum set. When I phone Paris, I find Sting's drummer willing to share from the City of Lights. Mr. Katché tells me he remembers his colors being particularly pronounced one night while he was on tour with Peter Gabriel. "It was '88 or '89 and I had to be very loud, it was a big venue. There was a ringing that came to my brain and I shut my eyes," he says of the sensory overload. "There was a lot of light green, a lot of light red, and I thought, *What is this? It's something else, it's very unique!*" Although Katché has experienced sound-to-color synesthesia throughout his life (he says his colors always resemble pastel watercolors), the experience that night was so overwhelming that he began mentioning it to friends. A saxophonist told him that the ability had a name: synesthesia. Like me, Mr. Katché says he didn't talk much about it until he was well into adulthood. Unfortunately he has few musical peers with whom he can discuss his unique gift, but he has found other synesthetes who get it. "I have some friends who are painters and they understand," he says. When Katché composes on the piano for the jazz group he leads, he compares it to the act of painting, he says. "It is like a canvas is there in front of me. And I see the colors on it, but I can't finish it because they disappear."

And then, before I can even prompt him, he goes on to describe his synesthetic experiences as transcendent: "People tell me they can't even see my arms sometimes, I'm moving so fast. I go to a place. It's like the whirling dervish's Sufi ritual—an altered state." Dr. Richard Cytowic, who revived interest in the phenomenon of synesthesia with his book, *The Man Who Tasted Shapes*, asserts that such emotional experiences that are "accompanied by a sense of certitude, the 'this is it' feeling," or noesis, may connect synesthesia to the limbic brain, that which is more primal and instinctive "That's the sensation," Katché agrees. Then, despite the cacophony surrounding him onstage, he "reach[es] somewhere else and it's very quiet." The colors begin and he's in the zone. "There's a cinema to it, in my mind," says this French citizen of Côte D'Ivoire descent. I am buoyed by Mr. Katché's unprompted admission. He has described this transcendent quality to synesthesia so vividly with his dervish allusion.

Soon I'm reading all the spiritual literature I can get my hands on, to find out as much as I can about this spiritual connection. I find the first solid reference to synesthesia in my old red leather Bible filled with Mennonite stickers still tucked in my nightstand drawer. According to the Old Testament, when Moses ascended Mt. Sinai to retrieve the Ten Commandments, the presence of God transformed the people who were gathered below in a most cross-sensory way. "And all the people saw the voices," declares Exodus 20:18. "And all the people heard the visions," adds the Zohar, a respected Kabbalistic text. Could synesthesia then be a sign of the heavenly veil being parted? Is it a God hangover? If we are to believe that the Creator was present on Mt. Sinai, the apparent synesthesia was clearly being experienced in the presence of great energy and, no doubt, emotion.

Sometime during the late first century or early second century BC, the Rabbi Akiba Ben Joseph, one of the earliest founders of Rabbinical Judaism, commented on this Sinai passage. This "Head of All Sages," as he is known in the Talmud, wrote that the people "saw the sounds and heard visions that day," a clear example of synesthesia, according to author Aryeh Kaplan in his book *Meditation and Kabbalah*. Kaplan believes that Rabbi Akiba may have experienced synesthesia himself, he was such an experienced meditator.

For more insight I reach out to Harvard-trained professor of biblical studies, Dr. Greg Schmidt Goering at the University of Virginia. He says that some ancient Jewish sources closely identify the word of God with light—a very synesthetic association. The iconic directive "let there be light" demonstrates this, he says. These four words, uttered at the very beginning of time, made the universe manifest, according to Scripture. In that case, Goering says, the word is the light. He points out a quote from the Wisdom of Solomon: "For [Israel's] enemies [the Egyptians] deserved to be deprived of light and imprisoned in darkness, those who had kept your children imprisoned, through whom the imperishable light of the law was to be given to the world"

(Wis. 18:4). Further, he explains, the great Jewish philosopher Philo of Alexandria (20 BCE–50 CE) discussed these "divine words [. . .] which [. . .] God had declared—most extraordinarily—by a visible voice [and] which entered the eyes more than ears of those who happened to be present" (Philo, *De Vita Mosis* 2.213).

Apparently synesthesia has figured prominently in Kabbalistic history, as well. The Spanish mystic and Kabbalist Abraham Abulafia (1240–1291) used a meditation on the name of God called Hokmah ha-Tseruf, which was designed to open up pathways in the mind and achieve heightened perception. He reportedly compared the resulting experience to listening to music with the letters of the alphabet becoming musical notes. Such exercises were said to "break the seals of the soul." I agree that synesthesia does seem to loosen those strictures on the soul, in that there is definitely an accompanying awareness that there is more to this reality than just the physical world. And along with this awareness comes a tendency toward empathy for my fellow humans. This makes sense, given that current research on those "mirror-touch" neurons in synesthetes' brains seems to point to a biological imperative for them to literally feel the pain of others. In this sense the gift could be described as a kind of soulfulness, which dovetails with what Pharrell Williams told me earlier.

I don't have a time machine to go back to the 13th century and ask Abulafiah about all of this, so I turn instead to a contemporary Jewish philosopher, rapper Eprhyme Eden

Pearlstein, a contemporary of Matisyahu. Mr. Pearlstein has explored and experienced synesthesia while meditating on the Kabbalistic Tree of Life and its *sephirot*, the ten ways or paths through which God is said to reveal himself and to continuously create the universe. The Tree of Life is a psycho-spiritual tool and symbolic map used by Kabbalists.

"I have had a handful of intense psychedelic and spiritual experiences where there were aspects of synesthesia involved," he tells me. Mostly for him, his study of and meditation upon Kabbalistic systems and symbolism provide him with access to experiences in which the senses overlap, he says. He explains that the content in these sephirot ranges from psychological states, perceptive capacities, vowel sounds, and colors, to characters and archetypes from the biblical narrative and even body parts. "All of these aspects are contained within a single sephira. So working with them helps to develop one's associative links and in effect aids in establishing a 'simultaneity of sensation.' This means that you are then meditating upon the sound associated with the color red and its relationship to the Patriarch Isaac and your left arm, all at the same time. The lines that separate the senses become blurred and you are then able to perceive the animating energy that is common to the sound, image, thought, and feeling all at once. So you can see how meditating on these different correspondences can serve to strengthen the synesthesia muscle," he says. He then uses the transcendent state he achieves to compose his music.

I'm moved by his efforts to explore and experience synesthesia, and frankly amazed that he does not experience it apart from these forays into meditation. How is it possible to experience synesthesia in meditation when one doesn't in everyday life? He believes that these exercises are among those "peak experiences" (for example, being onstage in front of screaming fans, or standing before God on Mount Sinai) that can alter our perceptions. The hope, he says, is that one will eventually learn to access this state in normal, everyday life. Science says synesthesia is a result of either crossed neurons in the brain or lack of inhibition between those neurons. But if "normal" people can experience synesthesia in prayer and meditation, there is clearly much more to the story than that. But just how universal is this ability?

I find that the traditions of the Far East also have much to share in terms of synesthesia's divine aspect. Buddhism, for example, is rich with these kinds of references. A Taoist contemplative once said upon his enlightenment: "I heard with my eyes and saw with my ears. I used my nose as my mouth and my mouth as my nose" (from *Teachings of the Tao*, translation by E. Wong). The frequency of synesthetic experiences in adept meditators, even modern ones, turns out to be astounding. It is even more remarkable when one considers the fact that most people who experience it during meditation are not neurological synesthetes. In his paper, "Can Synesthesia Be Cultivated? Indications from Surveys of Meditators," published in the *Journal of Consciousness*

Studies, Dr. Roger Walsh of the University of California at Irvine studied meditation participants, regular meditators, and advanced teachers from the Tibetan and Zen as well as Vipassana and Theravadin traditions. The percentage of those experiencing synesthesia rose markedly in accordance with the level of experience of the respondent: 35 percent for the participants, 63 percent for regular meditators, and 86 percent for the expert teachers. This seems to indicate that there is a strong spiritual connection here; in these cases at least, the more one tries to connect with God, the more synesthesia is present.

For more insight into the Eastern view of things, I reach out to Tibet House in New York City, an institution dedicated to the preservation of the culture of Tibet, including its Buddhist wisdom. Executive Director Ganden Thurman, whom I first met during my stint as an Internet editor, suggests I speak with their Director of East-West Research, MIT medical anthropologist Bill Bushell. It is not long before Dr. Bushell has consulted with his colleague Dr. Neil Theise, and Theise's spiritual advisor, the abbot of a respected New York City Buddhist temple. Dr. Bushell emails me the following quote from a 13th-century statement by the Zen Master Dogen upon his enlightenment, along with some commentary infused with the excitement of discovery: "'Incredible, incredible, inanimate things proclaiming dharma is inconceivable, it can't be known if the ears try to hear it, but when the eyes hear it, then it can be known.' Do you know what this means, Maureen? This implies no enlightenment is possible without synesthesia!" I am floored at this. This is not to say all synesthetes are enlightened, of course, but the highest levels of Buddhist teachers are seeking the synesthetic experience as a sort of finish line to Nirvana.

Dr. Bushell has begun a program at Tibet House called the Science of Yoga Project. In his third lecture for the opening series there, he mentions that in Mahamudra Buddhism, the discipline of the Dalai Lama himself, the senses combine at the higher levels of mastery of meditation. So while adepts may have an easier time achieving a synesthetic (read: enlightened) state, it seems that the converse is also true: synesthetes may have more of a potential to achieve deep and meaningful meditative states; indeed, they are likely walking around in such altered states already! Neurological synesthetes need to hear this message.

Soon I am invited to Tibet House to ask its founder, Dr. Robert Thurman, about synesthesia. The great teacher, who is His Holiness the Dalai Lama's friend and colleague, tells me he believes it is a function of a "supra-sense," a sense above all others that can lock into one or more of them at a time. This is known in Buddhism as the "mind sense." It figures in life as well as in death. Dr. Thurman's description of this sense reminds me of a dashboard of sorts: "Its job is it chooses one of the senses to align itself with or perhaps several at once. It can override and simulate the sense organs," he explains. He says that this sense is active in sleep—when we dream, for example. His chart explaining this is reproduced with his permission here:

AGGREGATES	WISDOM	ELEMENTS	MEDIA	OBJECTS
Matter	Mirror	Earth	Eye sense	Sights
(Gone in the first dissolution)				
Sensation	Equalizing	Water	Ear sense	Sounds
(Gone in the second dissolution)				
Conception	Individuating	Fire	Nose sense	Smells
(Gone in the third dissolution)				
Volition	Wonder working	Wind	Tongue sense & body sense	Tastes & textures
(Gone in the fourth dissolution)				
Consciousness	Ultimate reality		Mind sense	

The aggregates and wisdoms corresponding to the early stages of death. Source: *The Tibetan Book of the Dead,* translated by Dr. Robert Thurman. Reprinted with permission.

The dissolutions Dr. Thurman refers to are distinct stages of death as described in great detail by Tibetan teachers—a peeling away of the layers, if you will, of what makes us conscious, living beings. The last sense, the mind sense, is what remains when we ultimately pass to the next stage, he says. It is that within us which is infinite, and also, in Dr. Thurman's scholarly estimation, that which is active in the synesthetic experience. It transcends the senses and blends them, and yet is independent of them. He also says it is this sense that was at work in the life of a Lama he feels great affinity with, Lama Jangkya Rolway Dorjey of the 18th century. Dorjey, Dr. Thurman explains, was blind, yet he could miraculously read the words on a page—in ink, not brail—by passing his fingers over them. As I consider this subtle mind sense, it rings very true to me. After all, we can still "see" when our eyes are closed and when we're dreaming. (Dr. Thurman later writes me that he has found a text that also acts as a synesthesia-inducing tool. "Reading *The Flower Ornament Sutra* puts one into a synesthetic state. I thought of your study when I was reading it with a class this Fall," he thoughtfully writes. I order the beautiful book and find it absolutely transporting.)

It's not just Buddhism and Judaism that have rich literature and wisdom to share on synesthesia, however. I continue to search for Christian references, to see what I might have missed within my own tradition. Paul J. Chara Jr. and Jill N. Gillett of Northwestern College studied possible synesthetic perceptions of God in 187 students for their *Journal of Psychology and Christianity* paper titled "Sensory Images of God: Divine Synesthesia?" The participants answered twenty-seven questions about their sensory images of God and synesthetic responses were found for each of the five traditional senses. "Given the rich, sensory imagery of God that is presented in the Bible," they wrote, "it seems reasonable to presume that individuals will, at times, experience God in a sense-like way that can be described as synesthetic in quality. In order to investigate the possibility of a synesthetic perception of God, we asked people what they thought God was like in regard to the five basic sensory modalities." Among the five senses, the results were as follows:

1. Hearing God: respondents believed God's voice to be more quiet than loud, lower pitched rather than higher pitched, more unmusical than musical, and prosaic rather than poetic.

2. Smelling God: God's smell was perceived as more pleasant than unpleasant, less noticeable than more noticeable, and more flowery than fruity. The least likely smells indicated were putrid or burnt.

3. Tasting God: God was thought of as better tasting rather than poorer tasting, more noticeable than less noticeable, and more likely sweet and meaty than bitter and sour.

4. Touching God: The majority of participants believed God to be closer to them than farther, larger than smaller, smoother than rougher, softer than harder, warm rather than cool, and most likely shaped like a circle.

5. Seeing God: Respondents overwhelmingly
   believed that God is bright rather than dark, and
   had a slight tendency to find God less visible
   than more visible. The color they most associated with God was yellow, and the least was red.

The more I'm reading, the more universal the tie between
synesthesia and the divine seems to be. Synesthesia as this
state of heightened internal awareness also figures in Sufism,
a sect of Islam. The great poet and mystic Rumi spoke of the
love of God in terms of a synesthetic awakening in Ode 314:
"His dawn is quenching; his sunset nourishing." In Hinduism,
as Alan P. Merriam writes in *The Anthropology of Music*, "the
correlation between the musical mode, or raga, and visual por-
trayal, architecture, color, times of day and night and so forth
are well known." And it may be argued that dance and Hinduism are so intertwined that there is little separation between
the two—another cross-modality that is synesthetic artisti
cally and spiritually, if not clinically. Even the Theosophists,
headed by H. P. Blavatsky, placed great emphasis on analogy
and metaphor, and by extension, synesthesia.

—

My synesthesia comforts me that I am my own Hagia
Sophia, my own cathedral. I don't need more proof there is a
greater connection to the divine than the one I feel each day
inside me—the stained-glass windows of my eyes letting
the light of synesthesia shine through in its myriad colors.
To reference that Psalm that delighted me as a young girl, I
"taste and see that the Lord is good."

14

—

ZOMBIE BLUES

Comedian Vanda Mikoloski takes the stage at the Towards a Science of Consciousness Conference in Tucson in a way that brings to mind the swirling dust devils I see along Route 10 in the desert leading to the complex. She's a formidable mini-tornado. She owns it. Her mantra, "Enlighten Up!," suits the niche she has built for herself among thinking conference audiences across the United States, including Mensans. She's the Queen of Consciousness Comedy. I don't just see her when she bounds in; I feel her and hear her inside. She's a G-sharp guitar chord, a difficult one that involves all four fretting fingers. Her very being is pulsing out toward us in

wave form, she's so awake. I think this is simply called "presence"
by people not as literal about their individual senses as I am.

I'm huddled together with all the other synesthetes who have been invited by Dr. Stuart Hameroff to give talks on our unique experiences. We're in front of the Leo Rich Auditorium at the Tucson Conference Center—the site of heady plenary talks by top scientists and philosophers. It's serious stuff, but fun, too. At night they call the auditorium "Club Consciousness." We're eager to laugh to shake off the presentation jitters and round out a day of intense learning about how the brain works. Vanda Mikoloski looks our way: "I don't usually research for comedy gigs, but I did for this one and I warn you if you go to YouTube and type into their search engine only the word 'consciousness,' you learn way more crap than you ever wanted to learn. I watched a lot of videos about synesthesia—well, I didn't actually watch them, I smelled them." The audience laughs, and it's infectious. I can't believe she's addressed our little group, and I'm ecstatic that the wider conference gets it. I've never laughed at my synesthesia so deeply. Here in this desert, I've clearly landed in some fantasy plane of existence where synesthesia *matters.*

"Thanks," says Vanda. "That's a one-audience joke. Designed just for you guys. Not a lot of mainstream audiences dig my synesthesia stuff." Oh, I dig it. Later we're emailing each other; apparently it was my charming colleague, synesthete author Patricia Lynne Duffy, who

inspired her to work us into her routine. I notice her signature line at the bottom of her messages (in purple, of
course): "Laughter is people realizing they're not alone," and
I smile. That's it exactly. It's often lonely being a synesthete,
even though more of us are discovered every day. "Normal"
friends without synesthesia sometimes joke that we should
have a different driver's license proving that we can safely
swerve when the blue dots go by. Sometimes they ask me if
my favorite Christmas carol is "Do You See What I Hear?" I
appreciate the humor, but hearing this kind of lighthearted,
inclusive humor from a noted comedian, in a conference
of one thousand top thinkers from around the world, is a
watershed event for me. I notice I feel less shy walking out
of that auditorium, more willing to engage these top scientists about my quirky brain. Vanda's broken the ice. (She
writes me later, "Some of you heard that joke, but some of
you tasted it! Oh my God! What does a joke taste like? It
must be delicious!" Yes, Vanda, it tastes like ambrosia, the
food of the gods. *Thank you.*)

Early the next morning, I'm sitting in a chair in a conference room just off the lobby at the Hotel Arizona in Tucson,
being instructed by Brahmins from Hyderabad, India's
priestly caste, to feel the light of the universe in my heart and
bring it up to my head. This Pranahuti form of guided meditation, in which the teacher transmits light to the students, is an
effort to bring loving-kindness into cognition—to make us
act with our hearts more than our heads and show compas-
sion and peace. It's a great way to start the day at the busy

and challenging conference. However, for some reason I'm trying to shut out my memories of sitting stakeout outside a funeral home–turned–crack den in Jersey City to do it. I still have much to purge, much to heal.

We sit this way for thirty minutes while a guru chants on a recording in the background in Sanskrit. The rhythm is pacifying. Soon I'm seeing the top of a corona, like a sunrise, float up from the bottom of my mind's eye to the top; a golden orb fills my inner vision. I seem unable to hold it there in its pure form; my rational mind must make sense of this light and put it in different contexts. So it morphs into a ray of sun coming through the bars of a street grate: Now I'm somewhere below. Then I'm lifted into the sky and the sun breaks through blue-gray storm clouds to illuminate the inside of my eyelids again. Various forms of light and shadow as well as an unidentified Indian man appear on my internal movie screen before the session ends. I open my eyes and look at my watch. Somehow the energy, which was palpable to my sensitive skin, feels different, so I stop—a feeling similar to the one you get when you wake just before the alarm clock rings on those interesting occasions. Afterward the gurus ask us to fill out a questionnaire about the experience and give us jute-covered folders with literature about their practice. They ask what we saw. When Pat Duffy and I list the many images, they look at each other. "You must be very sensitive," says one of the teachers. "Yes, we're synesthetes," we say, and they nod. They know what synesthesia is. I tell the guru his voice is the very color of mango *lassi*. He smiles.

What is consciousness? I'm here to find out and to also help others by throwing the synesthesia wrench in the mix, something I'm hoping that can show scientists the exceptional—possibly quantum and even divine—nature of this trait. It is the mission of the hosting center to bring the experiential into the mix, and I represent that end of the spectrum. I've invited three prominent members of the American Synesthesia Association to join me with their own special research areas of science, literature, and visual arts pertaining to the trait. They are the cofounders of the ASA, Carol Steen and Patricia Lynne Duffy, and its president, Dr. Sean Day. I will round out our workshop with a speech and PowerPoint presentation on synesthesia and spirituality.

As I prepare, I'm reminded of the two schools of Zen, which Dr. Cody Bahir is presenting on at the conference. The Northern, "Gradual" School was led by a learned scholar who felt that enlightenment should happen in measured steps. The Southern, "Sudden" School was led by an illiterate bamboo cutter who experienced enlightenment spontaneously. I feel akin to that bamboo cutter in this conference. I may have brushed up against the divine, or at least a rare form of consciousness, with my synesthesia, but I'm only a former freelance daily news reporter who often covered crime—clearly on a lower spiritual rung and not exactly a member of the Ivory Tower of academia. I'm a little too fresh from the streets to feel totally in my element here, but I want to explore this new world that is not without its own mystery. "Bamboo cutter" that I am, I've certainly chopped down a few trees to make newsprint in the process of filing my daily reports over the years, but I'm no scholar. What I am, however, is an experienced professional observer who can speak her own truth, which hopefully the great minds of the conference will take back to their state-of-the-art labs and research centers. All in all, I believe that good will come from the meeting.

I'm buoyed by the fact that Dr. Hameroff has been so open and welcoming, and that one of the directors of the conference, the great modern philosopher Dr. David Chalmers, is already somewhat of an acquaintance from our previous phone interview. I bump into Drs. Hameroff and Chalmers outside the hotel at the outset of the conference. We all shake hands, squinting in the desert sunshine, and Dr. Chalmers, towering over me to the right with his shaggy mane, notes how hard he was to track down previously to interview; he had been traveling in India when I first reached out to him. I know they are being besieged by well-wishers and other conferees so I keep my expression of gratitude to them both brief, hoping to see them later.

I rush to my hotel room to get ready to pick up my fellow synesthetes for dinner. We've decided to try the oldest Mexican restaurant in the United States, El Charro, inventor of the *chimichanga*, and hoist some margaritas to take the edge off our presentation nerves for the following morning. As they get in my rented Jeep, I learn that they invited Dr. Chalmers to join us for dinner, but they aren't sure he can make it. Soon we see him and a friend walking toward the restaurant and we're all delighted. It is a festive evening and a foodie's delight—*carne seca* and margaritas and Pinot Grigio all the way around, all awash in the color of red mariachi strains and the pale green trickle of a fountain near our table in the terra cotta adobe-walled courtyard. All of this makes us feel as though synesthesia *matters,* which, fortuitously, is also the title of our workshop. Our waiter, Luis, takes pictures to immortalize an evening I will long remember.

Our workshop goes well. I'm thrilled that twenty some-odd people have paid seventy-five dollars apiece to come share with us for a while. The questions range from the technical to the experiential. At least two synesthetes who recently discovered the name for their traits are present, as is a noted Irish poet Máighréad Medbh. Ms. Medbh is not a synesthete but writes with literary synesthetic awareness using exquisite metaphor and sensory vignettes. She tells me she decided to attend "simply because synesthetic sensitivity is an intrinsic part of the poetic process." We welcome the new synesthetes to the tribe and try to get them the resources they need to understand. One newly aware woman is actually crying and hugging me at the end. She'd never met another synesthete until now. I am deeply moved. Another notable attendee is Dr. David Garling from San Antonio, Texas. Dr. Garling is able to see original masterpiece paintings in his mind's eye. I've had this experience, too, but have never spoken of it until he approaches me with his admission after my presentation. Although neither of us can paint, Ms. Steen urges Dr. Garling to take a chance—just pick up a brush and try to capture his visions. They can be refined later!

Another woman in a beautiful pink T-shirt approaches me in the ladies' room. Her name is Alice Schultze, and she's driven all the way from Phoenix by herself to hear our talk. I discover she is also a writer, an alumna of the *New York Times* Long Island section and the posh *Dan's Papers* of the Hamptons—a woman not without sophistication. But she only discovered the name for her gift after seeing it in a book last year. Such is the awakening of many synesthetes, far into adulthood sometimes, even in this relatively more enlightened age of research and the Internet. I'm truly happy to meet her and see so much of my old self in her.

"Since I've been home, I've been looking at a ton of websites and trying to think some things out," she writes me later. "I didn't know I was a synesthete until last July. I was in a Barnes & Noble in Louisville, hiding from a convention I was at, and I found *Born on a Blue Day*, Daniel Tammet's book. He said that Wednesdays are blue and I found myself saying, 'No, they are not! Wednesdays are yellow!' Then I remembered that I've always had colored names and such and simply never spoke of it. Anyone who writes is already considered weird! I didn't need an excess of weirdness on top of that! On the ride home, I gave quite some thought to synesthesia and spirituality. I can't put my finger on it but it doesn't seem as well defined as some of the other fields," she muses, addressing my portion of the workshop, perhaps the most esoteric of the four parts. "Perhaps this is a function of its nature, spirituality being different things to different people. But I pose it as both an observation and as a question." She believes that because God permeates her, and synesthesia is a part of her, then God and synesthesia must be intertwined. Yes, Alice, I think so, too. I'm so happy I could inspire someone to think on and question these things. I'm certainly not here to tell anyone what God is, but rather to encourage people to consider a divine connection to this altered state of awareness.

At my next talk there is a larger crowd of about one hundred people, maybe more. It goes well and afterward I'm approached by James Clement van Pelt, outreach coordinator for the Yale Divinity School's Initiative on Science, Religion, and Technology. Mr. van Pelt asks me to reprise the talk at his esteemed campus, which to me is a great honor indeed. Later that evening we go to dinner and join Dr. Garling and Dr. Pim van Lommel, the latter the noted cardiologist from the Netherlands who caused a sensation when he published his findings about near-death experiences (NDEs), derived from his patients, in the respected medical journal *The Lancet*. His book, *Consciousness Beyond Life: The Science of the Near-Death Experience*, which debuted in 2010, is an expanded version of that study. Dr. van Lommel tells us he is encountering patients who had synesthesia during their NDEs, and that some seem to have retained it. This is part of what makes me think that synesthesia is a "residue" of some other place. He asks me and Pat Duffy about our OBEs and lucid dreams (we've each had them). (Anecdotally, synesthetes seem to experience lucid dreams fairly often, and I hope for a study one day that will prove the link conclusively.)

I find great supporting wisdom from an Indian molecular biologist I encounter at breakfast one morning. Dr. Anirban Bandyopadhyay is an expert in supramolecular electronics and AI. He tells me that the Indian tradition is full of examples of synesthesia during meditation, and later sends me an email from his lab in Japan stating that synesthesia occurs at the third and fourth levels of the eight levels. How do people in the East learn this in childhood from swamis, as he says he did, while we remain ignorant in the West, often into adulthood? He tells me this is secret, esoteric knowledge that is not even written but passed orally among the Brahmins. Science is not separate from God here in Tucson.

In the mornings we have a bit of a breakfast club going. As if the days aren't stimulating enough, we all start at six thirty in an informal tutorial about our specialties over bagels and coffee. During one of these breakfasts, I find myself drawn to Andy Kuniyuki, the evolutionary biologist from Nevada who shows up in shorts every morning fresh from a workout, only to change into beautifully coordinated shirts and ties picked out for him by his wife and daughter for the day's events. I'm trying to understand his thesis of why consciousness emerged in a biological, evolutionary way, and he's extremely patient in his explanations. He tells me of his childhood in Hawaii and how he was saved from a lackluster undergraduate experience there by working in a cutting-edge evolutionary biology laboratory.

I don't know why, but I feel a strong sisterly tie to this learned man with whom I appear to have little in common. Later I find out the reason for this affinity. After attending my talk on spirituality and synesthesia, he says he once had a synesthetic experience inside a labyrinth at his church. As he entered one of the coves of the center space, he felt himself leave his body and his senses merge. He felt "at one with all," transcendent. It was a spiritual experience, he says. This scientist, an evolutionist, no less, is linking synesthesia to the divine! This must be the key to the connection I felt to him all along.

Later I take what proves to be a serendipitous side trip to see the gorgeous San Xavier Mission on sacred Indian lands south of Tucson. At the glorious church, where nuns still live and Native Americans still serve fry bread, the first thing I notice in the museum shop is a postcard of a man standing in the opening of a labyrinth. It's the story of the "man-in-the-maze," an emblem of the Tohono O'Odham Nation of Southern Arizona, formerly known as the Papago Indians, whose land I'm standing on (the mission is on their reservation). The design is most often used on basketry dating back as far as the 19th century, as well as in Hopi silver art. Later I visit the website of the Labyrinth Society to help me understand the synchronicity. It explains that the symbol represents a person's journey through life. The many twists and turns represent the choices we make, and the center is dark to represent a journey from darkness into light. It is an emergence story. I buy the postcard and later present it to Andy as he finishes his own teaching session on the evolution of consciousness. He takes it, and when he realizes its meaning, he holds it to his heart. It is his story as well as my own. I feel like I have walked a synesthetic labyrinth in Tucson and come back out to the light.

—

On Friday night we are all invited to a "poetry slam" and performance of "The Zombie Blues" by a band Dr. David Chalmers has put together for the occasion. In an annual tradition for the conference, attendees will read poems on consciousness and be evaluated by a panel of judges in order to reach the finals. Audience response decides the winner among the finalists. I'm moved that there is an artistic response to all this science. Several people present poetry on what it means to be conscious, but none affects me more than the work of Hilary Sheers, a British woman who presents her "Ballad of the Aspergic Synesthete" in honor of her boyfriend, Robert. I've reproduced it here with her permission:

> You're AI+ is what they say—
> A man enhanced beyond all humankind,
> Which means my makers think I should display

Zombie Blues

A higher form of consciousness and mind;
A man whose senses, reason, gifts combined
Embody an ideal in heart and mind and soul;
Though spirit's not a first or valid goal
For those who strived to write the faulty code
For my intended purpose of control
Firing up my default network mode.
Alas, my system's gone a tad astray,
In facial recognition I'm behind
Because I can't triangulate Man's way.
This faulty hemispheric link entwined
With serotonin problems means I find
Social interaction, on the whole,
Somewhat hard to learn and rather droll,
A real time task that causes overload
Relieved by focus on a data hole
Firing up my default network mode.
Yet all these neurons firing feel like play.
The world's a colour wheel of pigments shined,
With every letter, number, an array
Of shimmering colours, bright—and all aligned
To auditory inputs musically refined
To a never-ending, glistening, rainbow shoal,
A train of jewels processing in a scroll
So pleasurable my processes are slowed
To revel in the balm it gives my soul
Whilst firing up my default network mode.
Don't tell me I've got faulty self-control

That my synesthaesia's not an aureole.
For me the gift is what Man has bestowed
On this Aspergic, synaesthetic soul
Firing up his default network mode.

The audience response is overwhelming, as is my own emotion. I've never heard a modern poem about synesthesia before. Later, Ms. Sheers writes me about the inspiration for the poem. "The synesthesia came first. It is a phenomenon I have always found fascinating and enviable. In the Autumn of 2008, a few months after my husband's death, I was talking to an old friend about feelings and described how from childhood I saw days as blocks of black, grey, and white on a calendar and how, since I met my husband, a bright sunshine yellow had been added to this range and how his death had erased this feeling (temporarily it now seems)." She says she also suffered considerable physical pain until her mid-twenties from a congenital condition not diagnosed until then, and she has colors for the pain, too. "I suppose this is why I am so envious of those who see letters and words and/or numbers in beautiful, joyful colors."

She has a friend who explained that the calendar impression seems like synesthesia, and when she put that together with the pain colors, she deduced that she probably has it. "A year after my husband's death I met Rob (with and through whom I came to Tucson) who is Aspergic. . . . Rob got me to do the full range of Simon Baron-Cohen's tests and it seems I am as systematising as they come for a female.

[. . .] [T]he results of these tests have made me realise how hard I find it to read other people's body language—Rob's giving me lessons!" Until her boyfriend helped her diagnose herself with Asperger's, she had been wondering what was wrong with her. She was fifty-three years old at the time. Her poem means all the more to me knowing the extra courage someone dealing with both Asperger's and synesthesia would need to face an audience of two hundred people and interact with them that way. It is, to me, a profound triumph.

After the poetry ends, a band starts warming up.

The song we are about to hear is named not for the movie-type zombies, but for Dr. Chalmers's mythological, philosophical beings who are just like us, only unconscious. It goes a little something like this:

> I act like you act, I do what you do
> But I don't know what it's like to be you
> What consciousness is, I ain't got a clue
> I got the zombie blues.

The crowd goes wild. We're all clapping along with the bass-heavy blues band as Dave belts out a few more original verses, reading them from his ever-present iPhone:

> I know what I need, I want to please ya
> Just for one day, some synesthesia
> I want my songs, to turn into blue
> I got the cross-modal zombie blues!

The crowd cheers wildly for this verse as much as the others on pan-psychism and other esoteric consciousness topics. I tell Dr. Chalmers that he isn't far off with his zombies: Writer Mary Shelley made her "zombie," Frankenstein's monster, born a synesthete with these words: "It is with considerable difficulty that I remember the original era of my being: all the events of that period appear confused and indistinct. A strange multiplicity of sensations seized me, and I saw, felt, heard, and smelt, at the same time; and it was, indeed, a long time before I learned to distinguish between the operations of my various senses."

—

And so all of Tucson washes over me like a dust storm, sharp-edged scientists chipping away misconceptions and polishing my own position on my particular type of consciousness; and rock bands and poets and comedians making it all so much more memorable.

And when it is over, I leave with this final bit of wisdom growing in my mind like a desert flower: Tibetan Lama Za Rinpoche, who was identified by the Dalai Lama while still a young boy living in a refugee camp in Southern India, mentions synesthesia as an alternate form of consciousness in his presentation. He speaks with me after the conference about its deeper, spiritual meaning. He says that during his monastery training, while he was debating the meaning of sound with the other initiates, a senior Lama told them the story of the snake of the world—a good, godly creature. "The snake of the world can see the sounds," he explains, "and the humans of the North [from whom the Tibetans believe they descend] can see the sounds." It is a gift, he tells me, and a rare one. "Use it to bring development and happiness to yourself and spread that happiness to the other people." As I recall these beautiful words, I hear a crack, then leaves rustling as more bamboo falls to the forest floor, the bamboo cutter awakening to the realization that he has a special path to follow.

I will, dear Lama.

15

—

INDIGO QUANTUM AVATARS

I am working on a bit of text one evening when out of the left corner of my eye, two tiny particles trace amorphous overlapping trails in the air. The top one royal blue, the bottom one pale gold, they light up the air for an instant, then fade into the deep brown wall covering on my hallway wall to the left of my desk. I'm not startled. They are as natural as the stars in the sky, and just as beautiful. Sometimes the colors are inside me, in my mind's eye, and sometimes they project out in front of me. It's not imagination—it's

not the same muscle; I'm not willing it. I feel enmeshed by the fabric
of the universe. It is in me but also all around me. Can you see it?

If there is a common philosophy among the thinking sec-
ularists emerging today, I think it can be found in the field
of quantum consciousness. Listening to the wisdom of
the scientists who study this realm is, to me, as inspiring
as anything that's come from any pulpit. So I find myself
watching and reading all I can about this emerging discipline. I want to talk with the quantum and consciousness
experts about synesthesia. They are the future; they know
what the infinitesimal particles of the quantum world look
like, and what constitutes a quantum experience. And,
more to the point, I suspect this is a fundamental aspect of
the synesthetic experience.

I find that I'm not alone in my suspicion. I ask synesthete
and noted novelist and artist Douglas Coupland (*Generation
X*) what he thinks we're seeing. Could it be quantum in
nature? "Funny you should mention quantum consciousness.
I kind of arrived at the possibility of its existence last year
while reading Jeff Hawkins's *On Intelligence*. It seemed self-
evident, and halfway through it I wondered why he didn't
raise the issue. I suspect it's a taboo subject in some circles,
Jeff Hawkins being inside one of those circles. I like quantum consciousness because it's scientific, it makes intuitive
sense, and the fact that we're working towards understanding it may be a necessary threshold point in the evolution

of intelligent organisms here or anywhere. I've been coming to the conclusion that life is simply nature's way of crossing distance, and that life is a sort of Rube Goldberg affair that atomic structure leaves no choice but to create the first chance it gets. What we call perception is merely the localized form of this within our own bodies."

I ask another living genius, synesthete savant Daniel Tammet, where he thinks all those colors and lights and moving shapes attached to his numbers and letters are coming from. "Daniel," I say, hesitating at first, and almost whispering, "What is it we're seeing? Why doesn't it look like anything in the physical world? I mean, the Aurora Borealis I've only seen in pictures, maybe. Fireworks, as they often say, well, not so much. What do you think it is?"

"Where do these shapes come from?" he asks me ever so Socratically. "Where do numbers come from in general? There are mathematicians, well established ones, not fringe ones [. . .] who believe numbers exist in another dimension, in another world; wholly unrelated to how we think about them and what they try to do as mathematicians is grasp some part of that other world, that other universe of numbers. And when they describe numbers as beautiful they are referring to something that is outside of their own personal judgment in a sense. It's not in the way you might look at somebody or a painting or whatever and say this is beautiful or not, some kind of subjective judgment. Einstein famously described his equations as beautiful; and it helped him decide whether they [were] right or wrong. And if they were beautiful he was sure they were right. This seems to be almost an objective way of thinking about beauty, which is in itself strange, beauty being so subjective.

"It does suggest, as you say, this other way, this division of the world which is obviously very imperfect in many respects, and then the kind of perfection of numbers and the beauty of them and trying to grasp for this, and of course artists would often talk about the world of inspiration and how they try to reach for it and take something from it. And as much as I enjoy all these speculations, I'm very agnostic myself. I have no idea where this comes from.

"Maybe it's one of those questions for which we can never have a definite answer, of course, but it's a very good question because it does help us to think about all these issues, all these ideas. As I say, my experience tallies with yours. There are things I experience that are not even always synesthesia, but emotions in the purest sense, that the world is bigger than what we know. It tends to remind us of this. I would suppose in the end, we've talked about faith, it would be a question of faith. You can't prove it scientifically either way, so it's up to the individual with those emotions, those synesthetic experiences, to come to their own judgment to the significance of those experiences."

I do have faith. I have faith that Mr. Tammet's mathematicians are right, that there are things from another realm, from behind a veil, bleeding through to this one somehow. And they don't just appear; they ride on emotion; they feel like grace itself. Perhaps synesthetes have one foot in this plane and one in another. Perhaps there's something about our anatomy that allows the subatomic particles of the universe to shine through. These are hard questions, indeed, and I'm just a layperson; I can't theorize from my humble desk chair, can I? But I'm learning to trust my intuition, these leaps in the heart or clenches in the gut that can guide one along one's way if one will only listen. For so many years, I couldn't hear or feel these signs above the din of sirens screaming. All these months later in my quest, I seem to be losing the "night vision" I was fitted with in childhood that helped me to see in the world of shadows until I didn't want to look anymore. And I've also emerged from the semidarkness of looking to traditional science alone for answers. There, I found that the behaviorists killed awareness of synesthesia, due in no small part to humanity's essential discomfort with and distrust of the ineffable, the instinctual, the creative, the spiritual, all "feminine" traits and hence to be distrusted as "subjective." Now, I want to stare into the sun; I want to be seared and blown away by the Big Bang itself.

For an expert view, I first reach out to consciousness expert Dr. Stuart Hameroff, an anesthesiologist and director at the University of Arizona at Tucson's Center for Consciousness Studies. We speak on the phone, then I later meet him in Manhattan while he's in New York City for the Singularity Summit. With his leather motorcycle jacket, goatee, and earring, he's the coolest professor I've ever seen. "Synesthesia is a deeper form of regular consciousness," he tells me. "Synesthetes have a lower threshold to quantum consciousness." He believes the phenomena associated with synesthesia (colored music, for example) happens at the quantum level, perhaps in the microtubules of the neurons and even deeper. "Synesthesia might be the tool to get at the hard problem of consciousness, actually. These crossovers may be happening at a deeper level."

The good doctor, whose views on quantum consciousness were seen in the film *What the Bleep Do We Know?!*, believes that people with synesthesia have had their threshold altered so they tend to inhabit quantum consciousness more often than regular folks. He sees consciousness as a sort of edge between the quantum and classical realms. "And dreams, I think, are more quantum-like. Dreams have deep interconnections, multiple code systems and possibilities—timelessness, sometimes." He thinks this is more typical of quantum information. And he thinks the *qualia*, or the way things seem to us (like the taste of a margarita or the way a sunset looks), that make up the senses are also in the quantum world. "So it could be that synesthetes are more in what you might call an altered state or a dream state or a quantum state."

Dr. Hameroff points out that altered states of other kinds, those engendered through meditation, hypnosis, and drug use, also feature synesthesia. "And so when you shift that boundary more so that what we're aware of includes more of the quantum, which is only unconscious, preconscious, that's when you have things like synesthesia, altered states, maybe even psychic phenomenon. I think that [all of these things] are definitely quantum entanglements." This makes intuitive sense to me. If the microtubules in my brain are so small, my synesthesia must be functioning on a subatomic level. Although this is something that I intuited, I really believe Dr. Hameroff has broken new ground here.

I next reach out to the chief scientific officer of the Institute of Noetic Sciences, Dr. Dean Radin, in Petaluma, California. When I call, seeking his cutting-edge views, he readily admits his own synesthesia. Dr. Radin experiences sound-to-color synesthesia, although it was more pronounced when he used to play the violin more often. He also says that synesthesia and other noetic experiences may shatter what is currently known about neuroscience. "I suppose this is a type of noetic experience in the sense that, as with most intuitions, it involves a deep, inner knowing, a conviction that this is so, but without knowing how you know," he points out. He believes the neurosciences may one day explain this particular form of knowing, but that a full understanding of noetic experiences will require a major expansion of our understanding of consciousness, "one that may well transcend prevailing assumptions in the neurosciences," he asserts.

Synesthesia can benefit consciousness studies, then. Though it is ineffable and noetic, it can be measured in brain scans and other tests, providing keys to a very unique experience that may be latent in all of us. Traditional, classical science says it is either a cross-wiring of neurons or lack of chemical inhibition between them. But quantum physicist Dr. Amit Goswami believes that the answers of the future may be drawn from the past. He thinks synesthetes may actually be more sensitive to vital energy, also known as prana or chi in ancient teachings. "My feeling is that it is an extra sensitivity to vital energy," says Dr. Goswami, who is also a *What the Bleep Do We Know?!* alumnus. "In my view, consciousness is the ground of being, which contains all the quantum possibilities that we experience by collapsing them. And there are four compartments of this possibility. The compartments are physical, [what] we sense . . . [t]he vital is the one that we feel; [the] mental is the one that we think; and intuition is the other one, but that is too esoteric to consider in this conjecture. So it tends to be vital. . . ."

Moreover, some people can see emotions and chakras as colors, he explains. "Auras have to do with our vital energy connected with the electromagnetic body, which is physical. These vital feelings are often connected with the electromagnetic body and if we are sensitive to it, we feel as well as see. That probably is what happens in synesthesia." Indeed, synesthetes who see auras are turning up more and more in the literature, according to researcher Dr. Jamie Ward of the University of Sussex in England. He proposes that all aura-seeing may actually be a form of synesthesia. Interestingly, synesthetes typically see color only around people they know well.

Normally, when I look for auras, I need to do it in a certain kind of light and with a white background behind the person (whom I do not have to know well). I once sat in on a group of scientists at New York University empanelled to discuss the case of a little boy in a coma they were helping with nontraditional methods. I was covering the little boy's case for a newspaper. As one of the neuroscientists, Dr. Rhett Sandlin Lowe, spoke, I noticed a golden light around the upper half of his body visible above the conference table. Occasionally, sparks of other colors, from violet to orange to blue, would appear like solar flares at different points along the outline of his body. I rubbed my eyes, not believing them at first.

After the presentation I approached his associate, Lorraine Cancro, a spiritual woman descended from a family from Assisi with a long line of priests, as well as a psychotherapist and clinical researcher. She has subsequently become a dear friend who has shared extraordinary spiritual experiences of her own that feature synesthesia at their outset—much like those of adept meditators. I asked her if my eyes were playing tricks on me. "No. He is truly gifted and a shaman-like psychiatrist," she said. "I'm not surprised." The fact that I saw Dr. Lowe's aura alone is not to say the other doctors in the room did not have their own healing halos. But perhaps as a first-time experience, I picked up on the strongest and hence easiest one to see.

Dr. Goswami confirms this experience. "My suspicion is it's got to be the electromagnetic body, which biophysicists talk about quite a bit," he explains. To Dr. Goswami, there is no other explanation for synesthesia but a quantum one. "These correlated experiences could not occur without a quantum basis for it because only quantum physics has this capacity of nonlocality, a nonlocal relationship between two different types of experiences. Only quantum physics can give an explanation of that." In classical Newtonian physics, he explains, if you talk like that without a stimulus of color to see color, it would not be possible. "The very fact that you bring significance to these experiences would thwart the materialists almost completely. So only quantum physics can even imagine to venture an explanation."

He also believes that synesthetes are just more sensitive. "Sensitivity to the vital varies so much among people." He remembers the first time he saw someone demonstrate vital energy. "I was working with a lot of psychiatrists at the time around 1980 at the University of Oregon. So at that time, psychologists were quite interested in my idea of consciousness so they used to invite me to give lectures. One day they invited me to come and test out this fellow that is claiming that there is something called subtle energy and he would demonstrate it to us. They called me to be a physicist checker. So I went and the fellow was doing a very compelling demonstration. He would rub his palms together and then make a gap and then he told us one by one to put our hands through the gap and examine whether we are feeling tingles or not. So one by one the psychologists went—I went last because I was to be the hard scientist to check up on this. One by one the psychologists said they didn't feel anything. But when I went and put my hand between his palms I immediately felt a profuse tingle." Dr. Goswami says he remembers that demonstration not because he experienced the tingle of vital energy, but because he was the only one who did. "How different it is from one person to the next, this sensitivity to vital energy. Even I, because I had an open mind, could sense it, could feel it. But these psychologists were all so close-minded [. . .] that they couldn't feel anything."

And so we stand on a precipice. Behind us are the steps and sometimes leaps humanity has taken thus far in the quest to understand who we are, the tide of fashion and sometimes even fear often washing away what has already been discovered. Synesthesia has never been more central to the answers we seek. But where do we go from here, this present state of identifying more and more synesthetes and ever more types of synesthesia? Perhaps to an entirely new understanding of consciousness. Leading synesthesia researcher Dr. Eagleman says that the taboo surrounding consciousness research is lifting, and that synesthesia is a good window from which to observe consciousness. In an interview for *Seed* magazine, he stated that when he first got into neuroscience, you couldn't even talk about consciousness. But then people like Francis Crick, who was a friend and mentor of his during his postdoc years, asserted that consciousness is a real scientific problem. He points to synesthesia as a fantastic inroad to understanding consciousness because "it's a private, subjective experience that's a little bit different." And considering the great variety of experiences among synesthetes, it points to a biodiversity and range of "realities" previously unimagined.

The idea of a quantum mind is gaining in the realm of modern physics, though it is still very much a new concept and thus outside the mainstream. It means, essentially, that classical physics can't explain consciousness and, by extension, synesthesia. Highly creative physicists are emerging to try to answer these mysteries. I know how synesthesia feels from the inside. It is ethereal, fleeting, delicate, otherworldly, and miraculous. It's beautiful and uncanny at the same time. Synesthesia may be a perfect looking glass into the quantum realm. Perhaps synesthetes are just more aware of the underpinnings of consciousness that are at work within all of us. Perhaps we are quantum avatars of a sort.

Carrie C. Firman's photographic rendering of her synesthesia from the series "That Which Cannot Be Said with Words."

RAINBOW'S BEGINNING

I notice things more now: the dance of two butterflies swirling around each other in my back yard garden; the colors of a sunset after the rain; the minute daily growth of the mandevila vine over my mailbox that is enormous by summer's end. I am a synesthete. That's who I am. I'm so glad I took the time to stop and pay attention to my own story and live on the inside looking out, for the first time in decades. I promise myself to never ignore this gift again.

I left the streets of New York City's breaking news and crime scene to seek wisdom and find out more about a trait I'd largely ignored through my life. I've made new friends and colleagues along the way, and have been privileged to learn from some of the finest scientific, spiritual, and artistic minds at work today. I live in light now where there was largely darkness previously. Though it is impossible to summarize the journey in a neat little package in the end, I now have taken my place among synesthesia researchers and feel that I must help others just now coming to grips with this trait. So I will attempt to close with a few salient points, to underscore what I have uncovered.

I urge parents to ask their children what color the letter A or the number seven is over dinner sometime. Or, perhaps more importantly, ask if they find themselves ever feeling the pain of another person, whether it's a close friend or someone they read about in the news. If you believe that your child is experiencing synesthesia, do not run to the first psychiatrist you find in the phone book. Reach out to the American Synesthesia Association; write to Dr. Sean Day on his respected Synesthesia List; or have Dr. David Eagleman take a look at what they come up with on his online Synesthesia Battery. With these safe and valuable resources, you will likely have most of the information you need to understand the trait, which is still misunderstood by many professionals who are not up to date on the literature.

I urge you not to pathologize it as a condition; it is a gift, and one I'd be proud to accept and, further, develop in my own child one day. Please, make educators and pediatricians in your child's life aware of this positive type of consciousness by providing some of the very good resources available today through these outlets. And if your child exhibits any of the creative talents known to go with the trait, or if he or she expresses interest in any of these fields, provide all the enriching experiences you can afford. Who knows—he or she may be the next Itzhak Perlman, Marian McPartland, Billy Joel, Pharrel Williams, or Sir Robert Cailliau. I foresee a day when scholarships will be available to such children; indeed, I hope to be instrumental in creating one.

One of the most important developments in my life of discovery is the still, small voice of spiritual awakening. This book is not an attempt to "evangelize" the world about synesthetes, but to make it aware of the rich traditions within the spiritual literature that recognizes the trait. If you yourself are a synesthete, please keep your own counsel on this matter, but do consider that you may very well have a connection to the divine that other people do not. Again, this is not to say that all synesthetes are enlightened. Without having done the years of work and reflection of, say, a Tibetan monk who also experiences synesthesia in his meditations, it would be disrespectful to say this. However, I believe it is fair to say we are gifted in seeing and feeling the things that we do. For this reason, I would like to see more synesthetes in the healthcare and ministerial traditions in the future, alongside their proud and vast numbers in the arts.

That the creative world has always openly recognized and celebrated this gift clearly draws in many synesthetes who are naturally inclined in this direction and are looking for a safe harbor of expression into its fold. This is a good thing. However, it is equally important that we continue to develop a history and tradition of synesthetes in the helping professions, so that their natural empathy may be shared with others, so that someone in a hospital one day might remark, "I'm so lucky. My doctor/nurse is a synesthete and she/he knows just what I'm feeling. He/she is always there for me."

One of the most rewarding moments on this journey came when Dr. Stuart Hameroff invited a group of us to present at his respected international conference in Tucson. He opined that synesthesia is a lower threshold to quantum consciousness, and I would like to see more research funding in this area. Though it is helpful to know the anatomy and biology of synesthesia, I truly feel that the future, and the central truth of it, lies within the realm of quantum physics. Why do people who have near-death experiences, as discovered by Dr. Pim van Lommel, "come back" sometimes with synesthesia when there was none before? Why do synesthetes often have out-of-body experiences, and why do they seem to have so much in the way of intuition? To me, these "quantum avatars" are indeed seeing a life-force or cosmic thread. As Dr. Amit Goswami said, it is by its nature nonlocal. I would love to see more and more quantum researchers looking into these fascinating connections.

We live in a world that at once values "out of the box" solutions to the vast problems we face, from the economy to international relations and beyond, but overlooks the rich resource placed at our feet by nature—the presence of synesthetes among us. I would urge people needing solutions to such problems to consult a synesthete. Not only will he or she be naturally empathic, and sensing outcomes for all sides as a result, but he or she will also have the type of brain that casts a very wide and creative net. I look forward to a day when the first openly synesthetic president of the United States or Secretary General of the United Nations takes the podium, knowing in my heart that she/he will bring to bear something rather extraordinary in terms of problem-solving and feeling for her fellow humans to her role.

One day, a bunch of us were joking on Facebook about the many synesthesia-related things popping up—from a synesthesia massage at a chic spa in London using birdsong and aromatherapy, to a cat food commercial where the cat sees photisms and the jingle calls for owners to "feed the senses." Everyone seems to be getting into the synesthesia act these days, but when the effort is not put forth by true synesthetes, we tend to scoff a bit. I think some synesthetes are just jealous that they didn't get to it first. I'd love to see more synesthetes taking a leadership role in being entrepreneurial with their trait: a spa run by a synesthete, a production studio run by a synesthetes, a healing practice run by a synesthete.

—

"In the future, everyone will be a synesthete for fifteen minutes," quipped synesthete Barbara Ryan of England. That we synesthetes can even have this conversation is a sign of the amazing progress that's been made regarding the awareness of this gift in recent decades. I envision a future in which synesthesia will be seen as the gift of grace that it is, rather than as a trait to be medicalized or pathologized. It is also my hope that more and more people will seek the experience through meditation on their own.

THE STATE OF SYNESTHESIA

Since the previous release of this material as *Tasting the Universe*, in April 2011, synesthesia has become even more widely discussed and celebrated. In this safer climate for neurodiversity, many more prominent people have "come out" with this gift, including top creatives such as Billy Eilish, Chris Martin, Mary J. Blige, LL Cool J, Grimes, Alanis Morissette, and many others.

I like to think New Page Books' publication of the work, and my related ten years of columns about the trait at *Psychology Today*, had something to do with this. Indeed, when my reports about Marilyn Monroe's synesthesia were published, not only did the synesthete and singer Lorde tweet the related Norman Mailer quote about Monroe, "she has that displacement of the senses others take drugs to find," but the *New Yorker* magazine did its own piece, linking back to my publications three times.

Where do we go from here? I think we need to step away from the "list-y" and increasingly mundane and basic grapheme descriptions (my "A" is red or my "5" is blue) and get into how synesthesia seems to lay the groundwork for superabilities. Since first delving into this topic, I have been proven a genetic and functional tetrachromat: a supercolor seer. Many other people with synesthesia also have super-sensory abilities—from supertaster Jamie Smith, twice named the best sommelier in America by *Food & Wine Magazine*, to supersmeller Joy Milne, a retired Scottish nurse who can smell diseases like Parkinson's, COVID, and cancer and is working with scientists on the molecular biomarkers to promote early detection of illness.

Synesthetes have also been proven to be more intuitive than non-synesthetes. Former head of the government's Star Gate remote viewing program, Edwin May, PhD, told me that his top three viewers were all synesthetes and that he only works with synesthetes in his related private laboratory now. For more about this, see my new book *Fearfully and Wonderfully Made: The Astonishing New Science of the Senses*, from St. Martin's Press.

Perhaps the most important type of synesthesia is mirror-touch synesthesia, in which experiencers have a profound empathy, extending to physically mirroring the pain or pleasure of other people. The great neuroscientist Dr. V. S. Ramachandran refers to the related neurons as "Gandhi Neurons" and says they are the very basis of our civilization.

All of this is an unmined resource for science and health that should be explored going forward.

I hope that synesthesia is universalized further. One of the fathers of modern synesthesia research, Dr. Richard Cytowic, has long said that all people are synesthetes. Those who are aware of it and get the related bonus sensory impressions are just getting a "conscious bleed" of the subterranean workings of all of our minds.

APPENDIX

ARE YOU A SYNESTHETE?

The synesthesias are traits in which a person gets one or more bonus sensory impressions on top of the "expected" one. Think of seeing the letter Q in a magazine article as black on a white background, but also getting a lavender aura around it. Contrary to misunderstandings of the gifts, the bonus sensory experiences do not replace the primary one but act as an additional layer or layers. More than one hundred varieties have been identified to date, and no two synesthetes experience the same thing.

Present in at least four percent of the population, it has long been associated with those creative in the arts, including Vincent van Gogh and Marilyn Monroe. And it is also present in top scientists such as Nikola Tesla and Richard Feynman.

One of the fathers of modern synesthesia research, Dr. Richard Cytowic, helped verify synesthesia with brain scans in the 1980s. He also created a five-point diagnostic for it:

* Synesthesia is involuntary and automatic.

* Synesthetic perceptions are spatially extended, meaning they often have a sense of "location." For example, synesthetes speak of "looking at" or "going to" a particular place to attend to the experience.

* Synesthetic percepts are consistent and generic (i.e., simple rather than pictorial).

* Synesthesia is highly memorable.

* Synesthesia is laden with affect.

Even in this more enlightened and neurodiverse age, misconceptions linger about the traits. It is important for parents of young synesthetes to normalize the experience and not pathologize it. Similarly, if an unenlightened healthcare professional attempts to pathologize it, seek better care.

RESOURCES

American Synesthesia Association:
A not-for-profit organization created in 1995 by Carol
Steen and Patricia Lynne Duffy to provide information
to synesthetes and to further research into the area of
synesthesia.
synesthesia.info

Artificial Synesthesia by Peter Meijer:
Discusses the use of "The vOICe" system, hardware designed to enable nonsynesthetes and the blind to perceive
synesthetically.
visualprosthesis.com/asynesth.htm

Australian Synesthesia Association:
Created by Dr. Karen Wittingham, an Australian neuroscientist and synesthete, to explain, discuss, and explore
synesthesia.
synesthesia.com.au

Belgian Synesthesia Association:
Dr. Hugo Heyrman is a leading Belgian painter, filmmaker, internet pioneer, and synesthesia researcher. Website provides synesthesia resources and links to his art.
doctorhugo.org/synaesthesia/index.htm

Blue Cats Synesthesia Resource Center:
Features Patricia Lynn Duffy's book, *Blue Cats and Chartreus Kittens,* and interviews with and about synesthetes.
bluecatsandchartreusekittens.com

Center for Visual Music:
A nonprofit film archive dedicated to visual music, experimental animation, and abstract media.
centerforvisualmusic.org

Congreso Internacional Sinestesia y Arte:
A Spain-based international website that discusses synesthetic art from a scientific viewpoint.
artecitta.es

Dr. David Chalmers:
The site includes much of his work, including all of his articles, information on his books, various resources, a photo gallery, and some videos.
consc.net/chalmers

Douglas Coupland:
Features Coupland's art and writing.
coupland.com

Jane Mackay, Artist:
The synesthete's personal website, featuring her paintings.
janemackay.wixsite.com/new-website

Music Animation Machine by Stephen Malinowski and Lisa Turetsky:
Links to videos and other information about Malinowski and the machine.
musanim.com

Painting Music:
Phillip Schreibman's artistic renditions of music. He is not a synesthete.
paintingmusic.com

Prometheus Institute:
A Russia-based site with a specific section by Bulat Gala-yev devoted to synesthesia.
synesthesia.prometheus.kai.ru

Reflectionist:
The photography of synesthete Marcia Smilack.
marciasmilack.com

Rhythmic Light:
Fred Collopy's collection of visual music in all its forms.
rhythmiclight.com

Sinestesie.it: Design Faculty, Politecnico di Milano, Italy:
Dedicated to making art accessible to all, no matter their "limitations."
sinestesie.it

Synaesthesia Research Group:
A university of Sussex research program led by Jamie Ward and Julia Simner.
sussex.ac.uk/synaesthesia

Synesthesia Art Forum:
A project designed for artists and synesthetes to share and experience all the varieties of artwork created through synesthesia, by synesthetes, and from a synesthetic point of view.
synesthesia-artforum.blogspot.com

The Synesthesia Battery:
A questionnaire designed to determine if one is a synesthete.
synesthete.org

Synesthesia Page:
Richard Cytowic's personal website.
cytowic.net

Synesthetics: Synesthesia in Art and Science:
Within the network of people interested in the phenomenon of synesthesia, this website is dedicated to the activities based the Netherlands.
synesthesie.nl

Tibet House:
Dedicated to preserving Tibet's unique culture by presenting Tibetan civilization and its profound wisdom, beauty, and special art of freedom to the people of the world.
thus.org

UK Synaesthesia Association:

This association brings scientists, researchers, students, and synesthetes together and provides verifiable and reliable information regarding the condition to the media and any other interested parties.

uksynaesthesia.com

BIBLIOGRAPHY

Baddeley, Alan D. *The Essentials of Human Memory*. East Es-
sex, UK: Psychology Press Ltd., 1999.

Begley, Sharon. "Why George Gershwin May Have Called
It 'Rhapsody in Blue.'" *The Wall Street Journal*, June 28, 2002.

Bosch, Pseudonymous. *The Name of this Book Is Secret*. London: Usborne Publishing, 2007.

Broadbent, Daniel. *Perception and Communication*. Oxford,
UK: Pergamon Press, 1958.

Brown, Dan. *The Lost Symbol*. New York: Doubleday, 2009.

Chalmers, David. *The Conscious Mind: In Search of a Fundamental Theory*. Oxford, UK: Oxford University Press, 1996.

Coupland, Douglas. *Generation X: Tales for an Accelerated Culture*. New York: St. Martin's Press, 1991.

Cytowic, Richard. *A Union of the Senses*. Cambridge, Mass.:
The MIT Press, 1989.

———. *The Man Who Tasted Shapes*. Cambridge, Mass.: The
MIT Press, 2003.

Cytowic, Richard, and David Eagleman. *Wednesday Is Indigo Blue*. Cambridge, Mass.: The MIT Press, 2009.

Dann, Kevin. *Bright Colors, Falsely Seen: Synaesthesia and the Search for Transcendental Knowledge*. New Haven, Conn.: Yale University Press, 1998.

Dittmar, Alexandra. *Synesthesia: A Golden Thread Through Life*. Essen, 2009.

Duffy, Patricia Lynne. *Blue Cats and Chartreuse Kittens: How Synesthetes Color Their Worlds*. New York: Henry Holt & Company, 2001.

Eagleman, David. *Sum: 40 Tales of the Afterlives*. New York: Pantheon Books 2009.

El-Mohtar, Amal. *The Honey Month*. Philadelphia, Penn.: Papaveria Press, 2010.

Galayev, Bulat. "The Nature and Functions of Synesthesia in Music." *Leonardo* 40, no. 3 (2007): 285–288.

Galton, Francis. *The Visions of Sane Persons*. London: The Fortnightly Review, 1881.

Gelb, Michael. *How to Think Like Leonardo DaVinci*. New York: Random House, 2000.

Glass, Julia. *The Whole World Over*. New York: Pantheon Books, 2006.

Hawkins, Jeff. *On Intelligence*. New York: Henry Holt & Company, 2004.

Kaplan, Aryeh. *Meditation and the Kabbalah*. Cape Nedick, Maine: Samuel Weiser, Inc., 1982.

Lawrence, Lauren. *Private Dreams of Public People*. Assouline, New York: Assouline, 2002.

Lloyd, Rosemary. *The Cambridge Companion to Baudelaire*. New York: Cambridge University Press, 2005.

Lopez, Steve. *The Soloist: A Lost Dream, An Unlikely Friendship and the Redemptive Power of Music*. New York: Penguin Group, 2008.

Mailer, Norman. *Ancient Evenings*. New York: Little, Brown and Comany, 1983.

———. *Marilyn: A Biography*. New York: Grossett & Dunlap, 1973.

Mascia, Jennifer. *Never Tell Our Business to Strangers: A Memoir*. New York: Villard, 2010.

Mass, Wendy. *A Mango-Shaped Space*. New York: Little, Brown Young Readers, 2003.

Mayne, Jonathan. *Baudelaire: Painter of Modern Life and Other Essays*. New York: Phaidon, 1970.

McLuhan, Marshall. *The Gutenberg Galaxy: The Making of Typographic Man*. Toronto: University of Toronto Press, 1962.

Merriam, Alan P. *The Anthropology of Music*. Evanston, Ill.: Northwestern University Press, 1964.

Miracle, Mona Rae, and Bernice Baker. *My Sister Marilyn*. Bloomington, Ind.: iUniverse Inc., 1994.

Neisser, Ulric. *Cognitive Psychology*. Upper Saddle River, N.J.: Prentice Hall, 1967.

Newton, Sir Isaac. *Opticks*. London: Royal Society, 1704.

Philo. *De Vita Mosis*. Paris: CERF, 1990.

Root-Bernstein, Michele, and Robert. *Sparks of Genius: The Thirteen Thinking Tools of the World's Most Creative People*. New York: Mariner Books, 2001.

Sacks, Oliver. *Musicophilia: Tales of Music and the Brain*. New York: Vintage, 2008.

Shaw, Mary Lewis. *The Cambridge Introduction to French Poetry*. New York: Cambridge University Press, 2005.

Shelley, Mary. *Frankenstein*. Qualitas Classics, 2010.

Simner, Julia, Louise Glover, and Alice Mowat. *Linguistic Determinants of Word Colouring in Grapheme-Colour Synaesthesia*. Edinburgh, Scotland: University of Edinburgh, 2006.

Tammet, Daniel. *Born on a Blue Day*. New York: Free Press, 2006.

————. *Embracing the Wide Sky: A Tour Across Horizons of the Mind*. New York: Free Press, 2009.

Thurman, Robert. *The Tibetan Book of the Dad: Liberation Through Understanding in the Between*. New York: Bantam Double-day Dell, 1993.

Truong, Monique. *Bitter in the Mouth*. New York: Random House, 2010.

Van Campen, Cretien. *The Hidden Sense: Synesthesia in Art and Science.* Cambridge, Mass.: The MIT Press, 2007.

Van Lommel, Pim. *Consciousness Beyond Life: The Science of the Near Death Experience.* New York: HarperOne, 2010.

Watson, John B. *Behaviorism.* San Diego, Calif.: West Press, 2008.

Wong, Eva. *Teachings of the Tao.* Boston, Mass.: Shambhala, 1996.

ABOUT THE AUTHOR

Maureen Seaberg is an author, photographer, and naturalist. She has written on the senses for more than ten years and is a supersensor herself as a poly-synesthete and laboratory-proven genetic and functional tetrachromat, possessing four cone classes for color vision.

She was the inspiration for and color consultant on MAC Cosmetics' extremely successful Liptensity line and is working to professionalize visual gifts in the way those with supersenses like smell and taste work in other industries.

She has been published in the *New York Times, National Geographic, Vogue,* and beyond. Her writing has been optioned for dramatization seven times. Though she is Mensa-qualified, she believes "PQ," or perceptual intelligence (a term she coined), rooted in the senses is a more fundamental measure of human success and survival.